EARTH SYNERGY:

Collaborative Pathways of Environment and Human Progress

(ISBN : 979-8-9950196-2-6)

Editors

Dr. Jay B. Pandya
Dr. Priya Lokare

BOOK SAGA®
PUBLICATIONS

Title: Earth Synergy: Collaborative Pathways of Environment and Human Progress

ISBN: 979-8-9950196-2-6

Editors: Dr. Jay B. Pandya, Dr. Priya Lokare

Published by: Book Saga Publications

Rahuri Factory, Rahuri, Ahmednagar, Maharashtra, India, 413706.

Website - www.booksagapublications.com

E-mail - contact@booksagapublications.com

Mobile- +918788403871

BOOK SAGA® PUBLICATIONS

Table of Contents

Chapter 1

Political Ecology: Governance, International Relations, and Sustainability

Tran Mai Uoc

(Ho Chi Minh University of Banking) & Nguyen Vo Nguyen Anh (University of Social Sciences and Humanities)

Abstract

Political ecology occupies a unique space at the intersection of political science, environmental studies, and international relations. It examines how power, governance structures, and global political arrangements shape the distribution of environmental resources, risks, and responsibilities. This chapter explores the foundational principles of political ecology, moving from local governance dynamics to the complex architecture of international environmental diplomacy. It discusses how states, multinational institutions, and civil society actors negotiate sustainability agendas within systems shaped by economic interests and historical inequities. Through an analysis of environmental governance frameworks, climate diplomacy, and sustainable development policy, the chapter demonstrates that ecological crises are never

Isbn: 979-8-9950196-2-6

purely natural events — they are deeply political phenomena requiring politically informed responses.

Keywords: Political ecology, environmental governance, international relations, sustainability, climate diplomacy, power, global institutions

Introduction

The relationship between politics and the natural environment has never been more consequential than it is today. Environmental degradation, climate change, resource scarcity, and biodiversity loss are not simply ecological problems — they are products of decisions made by governments, corporations, and international institutions operating within specific structures of power and interest. Political ecology, as a discipline, provides the conceptual tools to understand these connections, asking not only what is happening to the natural world but who benefits, who suffers, and whose decisions are shaping these outcomes.

Keohane and Victor (2011)[1] argued that international environmental governance operates less as a single unified regime and more as a complex of overlapping, fragmented institutional arrangements. This observation captures the central tension at the heart of global sustainability governance: the need for coordinated collective action in the face of deeply fragmented political systems. Understanding this tension is essential for any serious engagement with the politics of sustainability.

1.1 Foundations of Political Ecology

Political ecology emerged in the 1970s and 1980s as scholars began to question purely biophysical explanations for environmental change. Blaikie and Brookfield (1987)[2] were among the

Isbn: 979-8-9950196-2-6

first to systematically demonstrate that land degradation was not simply the result of poor farming practices or natural causes, but was deeply rooted in political-economic structures — including land tenure systems, market pressures, and colonial legacies — that constrained the choices available to local communities.

The discipline rests on several core principles:

- Environmental problems are inseparable from social, economic, and political contexts
- Access to and control over natural resources reflects existing power inequalities
- Scale matters — local environmental outcomes are shaped by national and global political forces
- Marginalized communities frequently bear the heaviest burdens of environmental degradation while contributing least to its causes

These principles have made political ecology a powerful lens for analyzing issues from deforestation in the Amazon to water scarcity in sub-Saharan Africa, where the proximate ecological cause is always embedded within a deeper political-economic story.

1.2 Governance Structures and Environmental Management

Environmental governance refers to the rules, institutions, norms, and decision-making processes through which societies manage their relationship with the natural environment. Effective governance is foundational to sustainability, yet it operates across

Isbn: 979-8-9950196-2-6

multiple and often competing scales — from village-level resource management committees to global climate treaty bodies.

Ostrom (1990)[3] demonstrated through her landmark research on common-pool resource governance that communities are capable of developing sophisticated self-governing institutions that manage shared resources sustainably, challenging the dominant narrative that collective resources inevitably suffer the "tragedy of the commons." Her work highlighted that governance effectiveness depends less on the type of institution — state, market, or community — and more on the specific design principles that make rules legitimate, enforceable, and adaptive.

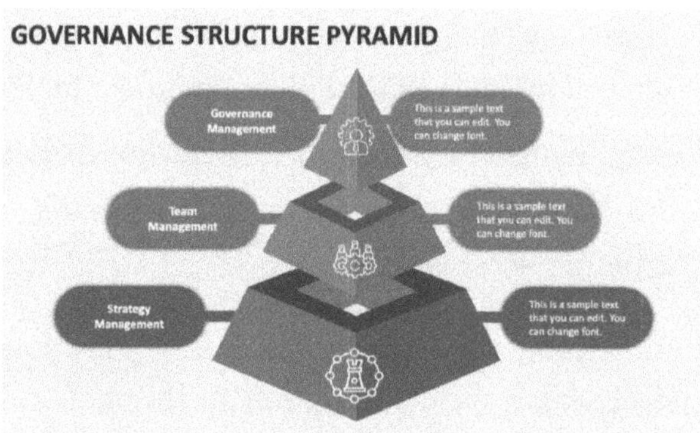

Figure 1: Multilevel Environmental Governance Framework:

At the national level, environmental governance is shaped by the strength of environmental institutions, the degree of democratic accountability, and the extent to which environmental ministries can resist capture by industrial and extractive interests. In many developing nations, governance capacity is constrained by limited financial resources, weak enforcement mechanisms, and the competing priorities of economic development.

1.3 International Relations and Environmental Diplomacy

The global nature of environmental challenges — particularly climate change — demands international cooperation on a scale rarely achieved in human history. Climate change does not respect national borders; its causes are globally distributed, as are its consequences, though in deeply unequal ways. This creates both the imperative for international coordination and the political difficulties that make such coordination so elusive.

Young (2011)[4] observed that the architecture of international environmental governance is characterized by institutional fragmentation, with hundreds of multilateral environmental agreements operating in parallel, often without effective mechanisms for coordination or mutual reinforcement. The Paris Agreement of 2015 represented a significant departure from this pattern, adopting a hybrid approach that combines legally binding procedural commitments with nationally determined substantive targets, allowing countries to set their own ambition levels while embedding them within a system of transparency and progressive review.

Environmental diplomacy is further complicated by the deep historical inequities between developed and developing nations. The principle of common but differentiated responsibilities, enshrined in the Rio Declaration of 1992 and reaffirmed in subsequent agreements, acknowledges that while all nations share responsibility for environmental stewardship, industrialized countries bear a greater historical responsibility for cumulative greenhouse gas emissions and should therefore shoulder a larger burden of mitigation and financial support.

Isbn: 979-8-9950196-2-6

1.4 Sustainability as a Political Project

Sustainability is often presented as a technical challenge — a matter of developing cleaner technologies, more efficient resource use, and better environmental management practices. Political ecology challenges this framing, insisting that sustainability is fundamentally a political project involving contested choices about values, priorities, and the distribution of costs and benefits.

The concept of sustainable development, as articulated in the Brundtland Report of 1987 and subsequently embedded in the United Nations Sustainable Development Goals (SDGs), attempts to reconcile economic growth with environmental protection and social equity. Critics within the political ecology tradition have questioned whether this reconciliation is genuinely possible within existing economic structures, arguing that the imperative of growth is itself a primary driver of environmental degradation.

Robbins (2012)[5] emphasized that political ecology must move beyond critique toward the identification of concrete alternatives — governance arrangements, economic structures, and social practices that genuinely place ecological limits at the center of decision-making rather than treating them as constraints to be managed around. This constructive agenda has become increasingly urgent as the timeframes for effective climate action narrow.

Conclusion

Political ecology reveals that environmental sustainability is never achieved through technical fixes alone. It requires governance systems capable of representing the interests of marginalized communities, international institutions capable of enforcing

Isbn: 979-8-9950196-2-6

collective commitments, and political cultures that recognize ecological limits as non-negotiable boundaries rather than obstacles to growth. The architecture of environmental governance — from local resource management to global climate diplomacy — must be continuously interrogated for the ways in which it reflects and reproduces existing power structures. Understanding the political dimensions of ecological crisis is the necessary first step toward addressing it effectively.

References

1. [1] Keohane, R. O., & Victor, D. G. (2011). The regime complex for climate change. *Perspectives on Politics*, ,(1)*9* 23–7. https://doi.org/10.1017/S1537592710004068
2. [2] Blaikie, P., & Brookfield, H. (1987). *Land degradation and society.* Methuen.
3. [3] Ostrom, E. (1990). *Governing the commons: The evolution of institutions for collective action.* Cambridge University Press.
4. [4] Young, O. R. (2011). Effectiveness of international environmental regimes: Existing knowledge, cutting-edge themes, and research strategies. *Proceedings of the National Academy of Sciences,* 19860–19853 ,(50)*108.*
5. [5] Robbins, P. (2012). *Political ecology: A critical introduction* (2nd ed.). Wiley-Blackwell.

Isbn: 979-8-9950196-2-6

Chapter 2

Environmental Economics and Policy: Balancing Growth and Sustainability

Dr. Jyotsna Gupta

Assistant professor
S. S. Law college Shahjahanpur, 242001
Email Id - jyotishubhgupta@gmail.com

Abstract

Environmental economics occupies the critical space where economic theory meets ecological reality. It asks a deceptively simple question: how can societies pursue economic prosperity without destroying the natural systems upon which all prosperity ultimately depends? This chapter examines the foundational principles of environmental economics, including market failures, externalities, and the valuation of natural capital, before turning to the major policy instruments available to governments seeking to align economic incentives with environmental objectives. It explores the tension between growth-oriented economic models and sustainability imperatives, and discusses how evolving frameworks such as green economics, circular economy models, and

Isbn: 979-8-9950196-2-6

ecological economics are challenging conventional assumptions about the relationship between economic activity and environmental health.

Keywords: Environmental economics, market failure, externalities, carbon pricing, green economy, sustainable development, environmental policy

Introduction

For most of economic history, the natural environment was treated as an inexhaustible source of inputs and an infinite sink for waste. Natural resources were assumed to be freely available, and the environmental consequences of production and consumption were either ignored or dismissed as beyond the scope of economic analysis. The emergence of environmental economics as a distinct field in the mid-twentieth century represented a fundamental challenge to these assumptions, insisting that ecological systems must be incorporated into economic analysis if markets are to function in ways that are genuinely sustainable over the long term.

Pigou's early work on externalities laid the groundwork for understanding why markets, left to themselves, systematically overproduce pollution and underprovide environmental protection. Subsequent generations of environmental economists have built on this foundation, developing increasingly sophisticated tools for measuring environmental costs, designing policy instruments, and evaluating the trade-offs between growth and sustainability. Tietenberg and Lewis (2018)[1] provide a comprehensive account of how this body of theory has translated into practical policy across a wide range of environmental domains.

2.1 Market Failures and Environmental Externalities

The central analytical insight of environmental economics is that environmental problems are, at their core, problems of market failure. Markets fail to deliver environmentally sustainable outcomes when the full social and ecological costs of economic activity are not reflected in market prices. When a factory discharges pollutants into a river, it imposes costs on downstream communities and ecosystems that are not captured in the price of the goods the factory produces. This gap between private cost and social cost is what economists call a **negative externality**.

Key sources of environmental market failure include:

- **Negative externalities:** Costs imposed on third parties and ecosystems that are not priced into market transactions
- **Public goods:** Environmental goods such as clean air, biodiversity, and stable climate that are non-excludable and non-rival, making it difficult for markets to provide them efficiently
- **Open-access resources:** Resources such as ocean fisheries that are accessible to all users but rivalrous in consumption, creating incentives for overexploitation
- **Information failures:** Consumers and producers frequently lack accurate information about the environmental consequences of their choices

Isbn: 979-8-9950196-2-6

The concept of **natural capital** — the stock of natural assets including soils, water, air, and living organisms that provide ecosystem services to human societies — has become increasingly central to environmental economics. Costanza et al. (1997)[2] produced a landmark estimate of the global economic value of ecosystem services, arguing that the services provided by natural systems — including climate regulation, water purification, pollination, and soil formation — represent economic values of enormous magnitude that are systematically excluded from conventional national accounts.

2.2 Policy Instruments for Environmental Management

Governments have developed a range of policy instruments to correct environmental market failures, each with distinct strengths and limitations. These instruments can be broadly categorized into regulatory approaches, market-based mechanisms, and voluntary schemes.

Regulatory Appraches (Command and Control)

Traditional environmental regulation sets legally binding standards for emissions, effluents, and resource use, enforced through monitoring and sanctions. While effective when properly enforced, regulatory approaches can be administratively costly, economically inflexible, and susceptible to regulatory capture by the industries they are meant to control.

Isbn: 979-8-9950196-2-6

Market-Based Instruments

- *Carbon taxes* place a direct price on greenhouse gas emissions, creating incentives for firms and individuals to reduce their carbon footprint. The Pigouvian principle underpins this approach: by making polluters pay the full social cost of their emissions, the tax internalizes the externality and aligns private incentives with social welfare.
- *Cap-and-trade systems* set a total limit on emissions and distribute or auction permits, allowing firms to buy and sell permits according to their abatement costs. This approach achieves environmental targets at least cost across the economy.
- *Payment for ecosystem services (PES)* schemes compensate landowners and communities for maintaining natural systems that provide services to others, such as watershed protection or carbon sequestration.

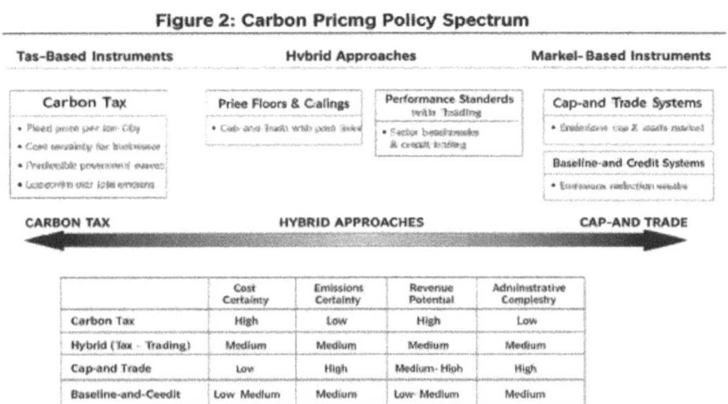

Figure 2: Carbon Pricing Policy Spectrum:

Isbn: 979-8-9950196-2-6

2.3 Balancing Growth and Sustainability

The relationship between economic growth and environmental sustainability is one of the most contested questions in environmental economics and policy. The Environmental Kuznets Curve (EKC) hypothesis suggests that environmental degradation initially increases as countries develop economically, but eventually declines as rising incomes generate demand for cleaner environments and the political capacity to regulate industry. However, the empirical evidence for the EKC is mixed, and many scholars argue that it reflects shifts of polluting industries to lower-income countries rather than genuine absolute decoupling of growth from environmental impact.

Daly (1996)[3] offered one of the most rigorous critiques of growth-centric economics, arguing that the macroeconomy operates within a finite biophysical system and that indefinite physical growth is thermodynamically impossible. His concept of the **steady-state economy** — one that maintains a stable throughput of materials and energy compatible with ecological limits — remains a foundational reference point for ecological economics and the degrowth movement.

More recently, the concept of the **circular economy** has gained significant policy traction as a framework for decoupling economic activity from resource consumption. Rather than the linear model of take-make-dispose, the circular economy seeks to keep materials in use for as long as possible through design for durability, reuse, repair, and recycling. Geissdoerfer et al. (2017)[4] examined the relationship between the circular economy and sustainability, finding important complementarities but also noting that circular strategies alone are insufficient without broader changes in production and consumption patterns.

2.4 Green Economy and Sustainable Development Policy

The green economy concept, promoted by the United Nations Environment Programme, frames environmental investment not as a cost but as an engine of economic growth, job creation, and poverty reduction. Green economy policies seek to redirect public and private investment toward renewable energy, sustainable agriculture, green infrastructure, and ecosystem restoration, arguing that the transition to sustainability creates economic opportunities that outweigh the costs of adjustment.

The Sustainable Development Goals (SDGs), adopted by the United Nations in 2015, provide the most ambitious multilateral framework for integrating environmental, social, and economic objectives. Goals relating to clean energy, sustainable cities, responsible consumption, climate action, and life on land and water all reflect the environmental economics insight that long-term prosperity depends on maintaining the natural systems that support economic activity. Sachs (2015)[5] provides a comprehensive analysis of how the SDG framework can be operationalized through integrated national planning and international cooperation.

Conclusion

Environmental economics provides an indispensable analytical framework for understanding why markets systematically undervalue the natural environment and for designing policies

Isbn: 979-8-9950196-2-6

that correct this failure. From carbon pricing to ecosystem service payments, the toolkit of environmental policy has become increasingly sophisticated, though its effective deployment remains dependent on political will and institutional capacity. The deeper challenge — whether economic systems can be genuinely reoriented around ecological limits rather than perpetual growth — remains unresolved, and its resolution will define the sustainability trajectory of human civilization in the twenty-first century.

References

1. [1] Tietenberg, T., & Lewis, L. (2018). *Environmental and natural resource economics* (11th ed.). Routledge.
2. [2] Costanza, R., et al. (1997). The value of the world›s ecosystem services and natural capital. *Nature, ,387* 260–253.
3. [3] Daly, H. E. (1996). *Beyond growth: The economics of sustainable development.* Beacon Press.
4. [4] Geissdoerfer, M., Savaget, P., Bocken, N. M. P., & Hultink, E. J. (2017). The circular economy: A new sustainability paradigm? *Journal of Cleaner Production,* 768–757 *,143.*
5. [5] Sachs, J. D. (2015). *The age of sustainable development.* Columbia University Press.

Earth Synergy: Collaborative Pathways of Environment and Human Progress
Isbn: 979-8-9950196-2-6

Chapter 3

Environment and Agriculture: Traditional Knowledge and Modern Practices

Dr. Nidhi Tripathi

Dr. Nidhi Tripathi Assistant Professor

Department of Zoology Swami Shukdevanand College Shahjahanpur U.P PIN code 242001

Email Id:- nidhi@sscollegespn.ac.in

Abstract

Agriculture is both one of the oldest human interactions with the natural environment and one of the most ecologically consequential. This chapter examines the relationship between environmental sustainability and agricultural practice, exploring the rich body of traditional ecological knowledge that indigenous and farming communities have developed over millennia alongside the transformative but often ecologically damaging innovations of modern industrial agriculture. It traces the rise of the Green Revolution and its contradictions, examines the ecological costs of chemical-intensive farming, and discusses the growing movement toward agroecology, organic farming, and climate-smart

Isbn: 979-8-9950196-2-6

agriculture as pathways toward systems that are both productive and environmentally sustainable. The chapter argues that the future of sustainable agriculture lies in a genuine dialogue between traditional wisdom and modern science.

Keywords: Traditional ecological knowledge, agroecology, Green Revolution, sustainable agriculture, biodiversity, soil health, climate-smart agriculture

Introduction

Agriculture feeds the world, but it also shapes the world in profound ways. Covering approximately half of the planet's habitable land surface, agricultural systems are among the most significant drivers of land-use change, biodiversity loss, freshwater consumption, and greenhouse gas emissions. At the same time, agriculture is the livelihood foundation for billions of people, particularly in the Global South, and the source of the food security upon which all human activity ultimately rests.

The history of agriculture is, in large part, a history of human communities learning to work with ecological systems — developing seed varieties adapted to local soils and climates, managing water through irrigation and rainwater harvesting, building soil fertility through composting and crop rotation, and managing pests through habitat manipulation and intercropping. This accumulated body of practice constitutes what scholars now call **traditional ecological knowledge (TEK)** — a dynamic, adaptive, and locally embedded system of environmental understanding that has sustained agricultural communities across diverse ecological contexts for thousands of years.

3.1 Traditional Ecological Knowledge in Agriculture

Traditional ecological knowledge in agriculture encompasses a vast range of practices, beliefs, and innovations developed by farming communities across cultures and ecosystems. It includes knowledge of seed saving and selection, water management, soil fertility maintenance, pest and disease management, and the integration of crops, livestock, and wild species within farming landscapes.

Berkes (2008)[1] defines TEK as a "cumulative body of knowledge, practice, and belief, evolving by adaptive processes and handed down through generations by cultural transmission, about the relationship of living beings (including humans) with one another and with their environment." This definition captures several critical features of TEK: it is cumulative, built over generations of careful observation; it is adaptive, continuously revised in response to changing ecological conditions; and it is embedded in cultural practices and social relationships that give it meaning and continuity.

Key examples of traditional agricultural knowledge with demonstrated environmental benefits include:

- **Shifting cultivation and fallow systems**, practiced across tropical regions, which allow soil fertility to recover and secondary forest vegetation to re-establish between cropping cycles

Isbn: 979-8-9950196-2-6

- **Terrace farming** in hilly and mountainous areas, which controls erosion, manages water, and creates microclimatic diversity
- **Intercropping and polyculture**, where multiple crop species are grown together, mimicking the structural diversity of natural ecosystems and providing natural pest suppression, nitrogen fixation, and microclimate regulation
- **Sacred groves and forest reserves**, maintained by community custom across South and Southeast Asia and sub-Saharan Africa, which preserve biodiversity and ecological functions within agricultural landscapes

3.2 The Green Revolution and Its Ecological Contradictions

The Green Revolution of the 1960s and 1970s represents the most dramatic transformation of agriculture in modern history. Driven by the development of high-yielding variety (HYV) seeds, particularly for wheat and rice, and accompanied by large-scale adoption of synthetic fertilizers, pesticides, and irrigation infrastructure, the Green Revolution dramatically increased crop yields across Asia and Latin America, averting the large-scale famines that had been predicted by demographers in the 1960s.

However, the ecological costs of this transformation were severe and are now well documented. Shiva (1991)[2] provided one of the earliest and most comprehensive critiques, documenting how the Green Revolution in Punjab replaced diverse traditional cropping systems with monocultures of HYV wheat and rice, depleting soil organic matter, lowering water tables through

Isbn: 979-8-9950196-2-6

intensive groundwater irrigation, contaminating soils and waterways with synthetic chemicals, and eroding the genetic diversity of traditional crop varieties that had been maintained by farmers over centuries.

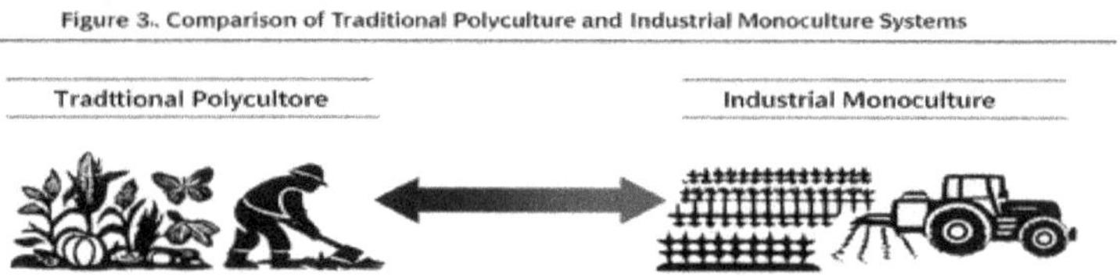

	Traditional Polyculture	Industrtal Monocuiture
Crop Diversity	High	Low
Soll Health	Improved	Degraded
Environmental Impact.	Low	High
Labor Intensity	High	Low
Economic Stability	Stable	Velatile

Figure 3: Comparison of Traditional Polyculture and Industrial Monoculture Systems:

3.3 Agroecology: Bridging Traditional Knowledge and Modern Science

Agroecology represents the most scientifically developed framework for integrating traditional ecological knowledge with contemporary agricultural science. It applies ecological principles to the design and management of agricultural systems, seeking to optimize the use of biological diversity and ecological processes rather than relying on external chemical inputs.

Gliessman (2015)[3] defines agroecology as “the application of ecological concepts and principles to the design and management of sustainable food systems,” encompassing not only the

Isbn: 979-8-9950196-2-6

biophysical dimensions of farming but also the social, economic, and cultural systems within which agriculture is embedded. Agroecological practices include:

- Diversified cropping systems that enhance biological pest control and reduce reliance on synthetic pesticides
- Legume integration and composting to build soil organic matter and reduce dependence on synthetic nitrogen fertilizers
- Water harvesting and conservation techniques adapted from traditional practices
- Seed sovereignty and the maintenance of locally adapted variety diversity

Altieri (1995)[4] demonstrated through extensive comparative research that agroecological systems, while often lower-yielding per unit area than optimally managed industrial monocultures, are more productive overall when the full range of outputs — food, fodder, fiber, fuel, medicinal plants — is accounted for, and are considerably more resilient to climate variability and pest pressure.

3.4 Climate-Smart Agriculture and Future Directions

Climate change is transforming the environmental context within which all agriculture operates, altering rainfall patterns, increasing temperature extremes, intensifying drought and flood events, and shifting the geographic ranges of pests and diseases. Climate-smart agriculture (CSA) has emerged as a policy framework seeking to address these challenges by simultaneously increasing agricultural productivity, enhancing resilience to climate variability, and reducing greenhouse gas emissions from agriculture.

The three pillars of CSA — productivity, adaptation, and mitigation — reflect an attempt to integrate the environmental and development dimensions of the agricultural challenge. However, Lipper et al. (2014)[5] caution that CSA remains contested terrain, with significant disagreements about the relative priority of these pillars and the degree to which existing industrial systems can be reformed versus the need for more fundamental transformation along agroecological lines.

Conclusion

The environmental future of agriculture depends on a genuine reconciliation between the productivity achievements of modern science and the ecological wisdom embedded in traditional farming knowledge. Neither uncritical reverence for tradition nor blind faith in technological innovation will produce agricultural systems capable of feeding a growing world population within planetary boundaries. Agroecology, with its commitment to both scientific rigor and respect for local knowledge, offers the most promising framework for this integration. The challenge is to scale these approaches without losing the diversity and adaptability that make them ecologically valuable.

References

1. [1] Berkes, F. (2008). *Sacred ecology* (2nd ed.). Routledge.

Isbn: 979-8-9950196-2-6

2. [2] Shiva, V. (1991). *The violence of the Green Revolution: Third world agriculture, ecology, and politics.* Zed Books.
3. [3] Gliessman, S. R. (2015). *Agroecology: The ecology of sustainable food systems* (3rd ed.). CRC Press.
4. [4] Altieri, M. A. (1995). *Agroecology: The science of sustainable agriculture* (2nd ed.). Westview Press.
5. [5] Lipper, L., et al. (2014). Climate-smart agriculture for food security. *Nature Climate Change,* 1072–1068 ,*4.*

Earth Synergy: Collaborative Pathways of Environment and Human Progress
Isbn: 979-8-9950196-2-6

Chapter 4

Environment and Law: A Global Perspective on Rights, Justice, and Sustainability

Dr. Ahinsa Pandey

Hyderabad
Email Id- at.tripathi30@gmail.com

Abstract

Environmental law represents one of the most rapidly evolving fields of legal scholarship and practice, responding to the unprecedented scale and complexity of ecological challenges in the twenty-first century. This chapter examines the development of environmental law from its origins in national pollution control legislation to the emergence of international environmental law as a distinct body of global governance. It explores foundational legal principles including the precautionary principle, the polluter pays principle, and intergenerational equity, and examines how these principles have been applied in landmark treaties, judicial decisions, and regulatory frameworks. The chapter also addresses the growing recognition of environmental rights as human rights, the emerging concept of rights of nature, and the role of

Isbn: 979-8-9950196-2-6

environmental justice in challenging the unequal distribution of ecological burdens across communities and nations.

Keywords: Environmental law, international environmental governance, precautionary principle, environmental rights, environmental justice, rights of nature, intergenerational equity

Introduction

Law is one of society's primary instruments for translating values into binding obligations and for allocating rights and responsibilities among competing interests. As environmental challenges have grown in scale and urgency, environmental law has expanded from a relatively narrow set of pollution control regulations into a vast, multi-layered system of rules governing the relationship between human societies and the natural world across local, national, and international scales.

The evolution of environmental law reflects a broader shift in how societies understand the environment — from a collection of resources to be managed for human benefit, to a complex system of interconnected processes that have intrinsic value and upon which all human activity depends. Birnie, Boyle, and Redgwell (2009)[1] trace this evolution in international environmental law, showing how each successive wave of treaty-making has expanded both the scope of legal protection and the circle of actors recognized as having rights and responsibilities in relation to the environment.

4.1 Foundational Principles of Environmental Law

Several core principles have come to define the normative architecture of modern environmental law, providing a basis for legal reasoning and policy design across diverse jurisdictional and cultural contexts.

The Precautionary Principle holds that where there are threats of serious or irreversible environmental damage, lack of full scientific certainty should not be used as a reason for postponing protective measures. This principle, enshrined in Principle 15 of the Rio Declaration on Environment and Development (1992), reflects the recognition that the pace of environmental change often outstrips the capacity of science to provide definitive causal proof, and that waiting for certainty may be catastrophic.

The Polluter Pays Principle establishes that those responsible for environmental damage should bear the costs of prevention, cleanup, and compensation. Originally developed within the OECD as an economic efficiency principle, it has been progressively incorporated into environmental law as both a cost allocation rule and a deterrence mechanism.

Intergenerational Equity, developed most fully by Weiss (1989)[2], holds that each generation holds the natural and cultural environment in trust for future generations and has both the right to benefit from it and the obligation to pass it on in no worse condition. This principle has significant implications for long-term environmental governance, supporting legal challenges to short-term decision-making that sacrifices future ecological capacity for present economic gain.

Isbn: 979-8-9950196-2-6

4.2 International Environmental Law and Treaty Architecture

The body of international environmental law has expanded dramatically since the Stockholm Conference of 1972, which established the foundational principle that states have the right to exploit their natural resources but also the responsibility to ensure that activities within their jurisdiction do not cause damage to the environment of other states or areas beyond national jurisdiction.

Subsequent decades have produced a complex architecture of multilateral environmental agreements covering climate change (the Paris Agreement), biodiversity (the Convention on Biological Diversity), ozone depletion (the Montreal Protocol), hazardous waste (the Basel Convention), and many other domains. The Montreal Protocol, which successfully phased out the production of ozone-depleting substances and has been credited with preventing catastrophic ozone layer depletion, is widely regarded as the most successful international environmental agreement in history. Velders et al. (2007)[3] documented the scale of this success, showing that without the Protocol, ozone depletion would have been dramatically more severe by the mid-twenty-first century.

Isbn: 979-8-9950196-2-6

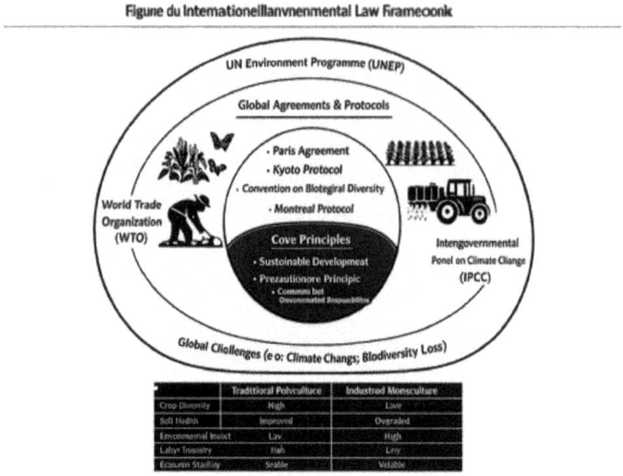

Figure 4: International Environmental Law Framework:

4.3 Environmental Rights and Environmental Justice

One of the most significant developments in contemporary environmental law is the progressive recognition of the right to a healthy environment as a fundamental human right. As of 2022, the United Nations General Assembly recognized this right in a landmark resolution, building on decades of advocacy by environmental and human rights organizations and codifying what many regional human rights systems had already recognized in their jurisprudence.

Environmental justice addresses the empirical reality that environmental burdens — pollution, industrial hazards, climate vulnerability, and degraded ecosystems — are not randomly distributed but fall disproportionately on communities that are already economically and politically marginalized. Bullard (1990)[4]

Isbn: 979-8-9950196-2-6

documented this pattern extensively in the United States context, demonstrating that race was the most significant predictor of proximity to hazardous waste facilities, a finding that has been replicated across diverse national contexts.

The emerging concept of **rights of nature** takes environmental legal protection one step further, extending legal personhood and rights to natural entities — rivers, forests, ecosystems — independent of their value to human beings. Ecuador's constitutional recognition of the Rights of Pacha Mama (Mother Earth) in 2008 was a landmark in this movement, followed by legal personhood granted to the Whanganui River in New Zealand and the Ganges and Yamuna rivers in India (though the Indian Supreme Court subsequently stayed the latter decision).

4.4 Environmental Law in the Context of Sustainability

The integration of environmental law with broader sustainability frameworks requires that legal systems move beyond reactive pollution control toward proactive governance of socio-ecological systems. Knox (2018)[5] argues that international human rights law and international environmental law are increasingly converging around a common framework of substantive and procedural rights — rights to information, participation, and access to justice in environmental decision-making — that provide citizens with the legal tools to hold governments and corporations accountable for sustainability commitments.

Isbn: 979-8-9950196-2-6

Conclusion

Environmental law has traveled a remarkable distance from its origins as a narrow set of pollution regulations to its current status as a complex, multi-layered global governance system engaging some of the most profound questions of justice, rights, and sustainability. The principles of precaution, polluter pays, and intergenerational equity provide a normative foundation for legal reasoning that takes ecological realities seriously. The progressive recognition of environmental rights and rights of nature reflects a deepening understanding that effective environmental governance requires not just better rules but a fundamental reorientation of legal systems toward the protection of ecological integrity as both a precondition for and a component of human well-being.

References

1. [1] Birnie, P., Boyle, A., & Redgwell, C. (2009). *International law and the environment* (3rd ed.). Oxford University Press.
2. [2] Weiss, E. B. (1989). *In fairness to future generations: International law, common patrimony, and intergenerational equity.* Transnational Publishers.
3. [3] Velders, G. J. M., et al. (2007). The importance of the Montreal Protocol in protecting climate. *Proceedings of the National Academy of Sciences,* 4819–4814 ,(12)*104.*
4. [4] Bullard, R. D. (1990). *Dumping in Dixie: Race, class, and environmental quality.* Westview Press.
5. [5] Knox, J. H. (2018). *Principles on human rights and the environment.* United Nations Human Rights Office.

Earth Synergy: Collaborative Pathways of Environment and Human Progress
Isbn: 979-8-9950196-2-6

Chapter 5

Environment Across Disciplines: A Global Interdisciplinary Perspective

Dr. Abha Dixit

Assistant Professor (Tier III)
Department of Genetics & Plant Breeding, Chandra Shekhar Azad University Of Agriculture & Technology
Address: CSAUAT Agricultural College Lane, Nawabganj, Kanpur, Uttar Pradesh 208002.
Email ID: abhadixit2012@gmail.com

Abstract

Environmental challenges are inherently interdisciplinary. They cannot be understood through the lens of any single academic discipline, nor can they be effectively addressed by any single professional domain. This chapter makes the case for a genuinely interdisciplinary approach to environmental understanding, tracing the historical development of environmental studies as a field that deliberately crosses disciplinary boundaries and examining the contributions of natural sciences, social sciences, humanities, and applied professional fields to a shared understanding of human-environment relationships. It explores

Isbn: 979-8-9950196-2-6

how interdisciplinary collaboration is structured in practice, what epistemological challenges it faces, and how it is being institutionalized in research universities, policy bodies, and international organizations. The chapter argues that the environmental crisis is, at its deepest level, a crisis of fragmented knowledge — and that overcoming it requires new modes of inquiry that are as integrative as the problems themselves.

Keywords: Interdisciplinarity, environmental studies, knowledge integration, transdisciplinarity, human-environment relationships, systems thinking, environmental humanities

Introduction

The environmental problems of the twenty-first century — climate change, biodiversity loss, pollution, land degradation, and freshwater scarcity — share a common characteristic: they are produced at the intersection of natural and human systems and cannot be fully understood or effectively addressed from within any single disciplinary framework. A biologist studying ecosystem function, an economist analyzing resource allocation, a political scientist examining governance failures, and a cultural anthropologist investigating the values and practices that shape environmental behavior are all studying different facets of the same complex reality.

The recognition of this complexity has driven the emergence of environmental studies as an explicitly interdisciplinary field, drawing on natural sciences, social sciences, humanities, engineering, and applied professional fields to develop integrative frameworks for understanding and managing human-environment relationships. Frodeman (2010)[1] describes this as a "post-disciplinary"

Isbn: 979-8-9950196-2-6

moment in the history of knowledge, where the most pressing intellectual and practical challenges demand forms of inquiry that traditional disciplinary structures were not designed to support.

5.1 The Natural Sciences Foundation

The natural sciences provide the empirical foundation for environmental understanding. Ecology, atmospheric science, hydrology, soil science, marine science, and conservation biology generate the data and models that describe the structure and functioning of natural systems and document the impacts of human activity upon them.

The concept of **planetary boundaries**, developed by Rockström et al. (2009)[2], represents one of the most influential recent contributions of natural science to environmental interdisciplinarity. By identifying nine Earth system processes for which thresholds exist beyond which human perturbation risks triggering large-scale abrupt or irreversible environmental changes, the planetary boundaries framework provides a biophysical foundation for discussions about safe operating spaces that necessarily engage economics, governance, ethics, and social science.

Key natural science disciplines contributing to environmental understanding include:

- Ecology — studying the relationships between organisms and their environments
- Climate science — modeling atmospheric and oceanic processes and their interactions with human systems

Isbn: 979-8-9950196-2-6

- Hydrology — analyzing water cycle dynamics and the impacts of land use and climate on freshwater availability
- Conservation biology — examining the conditions required to maintain biodiversity across fragmented and disturbed landscapes

5.2 Social Sciences and the Human Dimensions of Environment

Understanding environmental change requires understanding human behavior, institutions, economics, and politics — the domain of social science. The social sciences contribute conceptual frameworks for analyzing why environmental problems arise, how they are perceived and valued by different communities, what governance arrangements can address them effectively, and how the costs and benefits of environmental change and environmental policy are distributed across society.

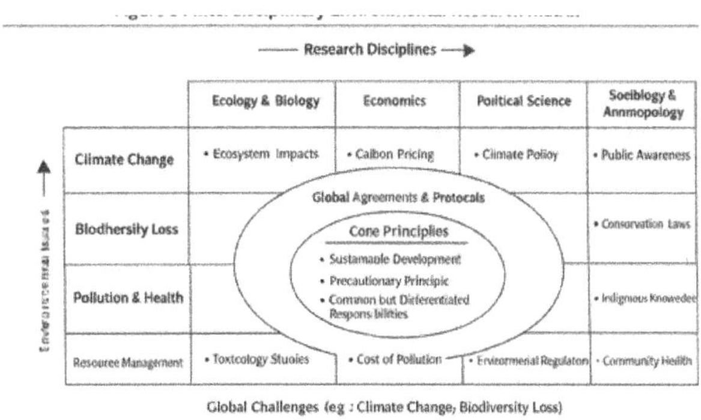

Figure 5: Interdisciplinary Environmental Research Matrix:

5.3 Humanities and Environmental Ethics

The humanities — including philosophy, history, literature, cultural studies, and religious studies — contribute dimensions of environmental understanding that the natural and social sciences alone cannot provide. Questions about how human cultures have understood and valued the natural world, what ethical obligations humans have toward non-human nature and future generations, and how narratives and representations of nature shape environmental behavior are fundamentally humanistic questions.

Environmental ethics, as a field of philosophical inquiry, examines the moral relationships between humans and the natural environment. It challenges the anthropocentric assumption that only human beings have intrinsic moral value, arguing for the extension of moral consideration to animals, ecosystems, and the biosphere as a whole. Naess (1973)[3] introduced the concept of **deep ecology**, which holds that the natural world has intrinsic value independent of its utility to human beings — a philosophical position with significant implications for environmental law, policy, and practice.

The **environmental humanities** have emerged as a distinct interdisciplinary field bringing together history, literature, philosophy, and cultural studies to examine the historical and cultural dimensions of human-environment relationships. Historians of the environment — including Cronon (1991)[4] — have documented how landscapes are simultaneously natural and cultural products, shaped by both ecological processes and the values, technologies, and economic systems of the societies that inhabit them.

Isbn: 979-8-9950196-2-6

5.4 Institutionalizing Interdisciplinarity

Despite the intellectual case for interdisciplinary environmental research, institutional structures within universities, research funding agencies, and policy bodies remain predominantly organized along disciplinary lines. This creates what Frodeman (2010)[1] calls the "disciplinary problem" — the tendency of knowledge institutions to reward depth within disciplines over integration across them, even when integration is what complex problems demand.

Responses to this challenge have included the establishment of interdisciplinary environmental research centers, the development of new graduate programs in sustainability science and environmental studies, and the promotion of transdisciplinary research methodologies that engage non-academic stakeholders — communities, policymakers, and practitioners — as co-producers of knowledge. Lang et al. (2012)[5] provide a framework for transdisciplinary research that explicitly addresses the challenge of integrating scientific knowledge with local and practitioner knowledge in ways that are both scientifically rigorous and practically useful.

Conclusion

The environmental crisis calls for a fundamental rethinking of how knowledge is organized and produced. No single discipline possesses the conceptual tools, empirical methods, or practical reach to address problems that are simultaneously ecological, economic, political, cultural, and ethical. Environmental studies, in its commitment to genuine interdisciplinarity, represents one of the most important intellectual developments of the contemporary

Isbn: 979-8-9950196-2-6

era — not merely as an academic exercise but as a practical necessity for developing the integrated understanding that effective environmental governance requires.

References

1. [1] Frodeman, R. (2010). *The Oxford handbook of interdisciplinarity.* Oxford University Press.
2. [2] Rockström, J., et al. (2009). A safe operating space for humanity. *Nature,* 475–472 *,461.*
3. [3] Naess, A. (1973). The shallow and the deep, long-range ecology movement. *Inquiry,* 100–95 ,(1)*16.*
4. [4] Cronon, W. (1991). *Nature's metropolis: Chicago and the Great West.* W. W. Norton.
5. [5] Lang, D. J., et al. (2012). Transdisciplinary research in sustainability science. *Sustainability Science,* 43–25 ,(1)7.

Isbn: 979-8-9950196-2-6

Chapter 6

Environment and Human Health: A Link Between Medicine, Pharmacy, and Ecology

Dr. Abha Dixit

Assistant Professor (Tier III)
Department of Genetics & Plant Breeding, Chandra Shekhar Azad University Of Agriculture & Technology
Address: CSAUAT Agricultural College Lane, Nawabganj, Kanpur, Uttar Pradesh 208002.
Email ID: abhadixit2012@gmail.com

Abstract

The relationship between environmental conditions and human health is one of the oldest and most fundamental concerns of medicine. Yet it has only been in recent decades, with the emergence of environmental health as a distinct scientific discipline and the growing recognition of ecological determinants of disease, that this relationship has been systematically incorporated into public health policy and medical practice. This chapter traces the connections between environmental quality and human health

Isbn: 979-8-9950196-2-6

across multiple domains — air and water pollution, chemical contamination, climate change, biodiversity loss, and ecosystem degradation — and examines how pharmacology and traditional medicine systems have historically drawn upon natural environments for therapeutic resources. It argues that human health and ecological health are inseparable, and that effective medicine must understand ecology, just as effective environmental management must understand human well-being.

Keywords: Environmental health, ecology and medicine, One Health, pharmacology, pollution and disease, climate change and health, traditional medicine

Introduction

Hippocrates, writing in the fifth century BCE, argued that physicians must understand the characteristics of the environment — the winds, the waters, and the soils — to understand the diseases that afflict the people who live within it. This intuition, ancient in origin, has been vindicated by modern environmental epidemiology, which has systematically demonstrated the profound influence of environmental conditions on patterns of human disease, disability, and mortality.

The World Health Organization estimates that approximately a quarter of global deaths are attributable to preventable environmental causes — including air pollution, unsafe water, inadequate sanitation, chemical exposures, and climate-related events. The burden of environmentally mediated disease falls most heavily on children, the elderly, and populations in low-income countries with limited access to healthcare and environmental protection. McMichael (2017)[1] provides a comprehensive account of how the broad sweep of human history — including the emergence of agriculture, urbanization, industrial production, and now global climate change — has repeatedly restructured the environmental determinants of population health.

6.1 Air Pollution and Respiratory Health

Air pollution is the most significant single environmental contributor to the global burden of disease. Both outdoor (ambient) and indoor air pollution contribute to respiratory and cardiovascular disease, cancer, neurological impairment, and premature mortality. The World Health Organization's revised air quality guidelines, published in 2021, significantly tightened recommended exposure limits for particulate matter, nitrogen dioxide, and ozone, reflecting the growing body of evidence that health effects occur at lower concentrations than previously understood.

Indoor air pollution, resulting primarily from the combustion of solid fuels — wood, charcoal, crop residues, and dung — for cooking and heating in poorly ventilated households, is estimated to be responsible for approximately 3.8 million premature deaths annually, predominantly among women and children in low-income countries. The transition to clean cooking fuels and technologies is therefore simultaneously an environmental, health, and gender equity intervention of enormous significance.

6.2 Water Quality and Waterborne Disease

Access to clean water is a fundamental determinant of human health, yet approximately two billion people worldwide continue to drink water contaminated with fecal matter, heavy metals, agricultural chemicals, or industrial pollutants. Waterborne diseases

Isbn: 979-8-9950196-2-6

— including cholera, typhoid, dysentery, and hepatitis A — remain major causes of child mortality in many parts of the world, accounting for hundreds of thousands of deaths annually.

The contamination of freshwater systems by agricultural runoff — particularly nitrogen and phosphorus from synthetic fertilizers — drives eutrophication of water bodies, stimulating harmful algal blooms that produce toxins dangerous to human health as well as devastating to aquatic ecosystems. Carpenter et al. (1998)[2] documented the scale of nonpoint source pollution from agriculture and its consequences for freshwater ecosystem health, arguing for integrated watershed management approaches that address land use, agricultural practice, and water quality simultaneously.

Figure 6: Environmental Pathways to Human Health Impacts:

6.3 Ecology, Biodiversity, and Medicine

The connection between biodiversity and medicine is deep and multidimensional. Natural environments have historically been the primary source of pharmaceutical compounds, and the

Isbn: 979-8-9950196-2-6

ongoing discovery of bioactive molecules from natural organisms continues to contribute to drug development across therapeutic categories. Taxol, derived from the Pacific yew tree, is one of the most widely used cancer chemotherapy agents. Artemisinin, derived from the plant *Artemisia annua* and used in the most effective treatments for malaria, was identified through systematic screening of traditional Chinese medicinal plants.

The **One Health** framework, which has gained significant traction in international health policy, recognizes that human health, animal health, and ecosystem health are inextricably linked and that effective responses to emerging infectious diseases — many of which originate at the human-animal-environment interface — require integrated approaches that cross the boundaries of human medicine, veterinary science, and ecology. Zinsstag et al. (2011)[3] provide a comprehensive account of the One Health concept and its implications for disease surveillance, control, and prevention.

The loss of biodiversity threatens this pharmacological resource base, as species that have not yet been scientifically investigated may harbor compounds of significant medicinal value. Preston (2011)[4] estimated that a substantial proportion of undiscovered natural compounds with potential pharmaceutical value may be lost to extinction before they are ever identified or studied.

6.4 Climate Change and Health

Climate change is increasingly recognized as one of the greatest threats to human health in the twenty-first century, operating through a complex web of direct and indirect pathways. Direct

Isbn: 979-8-9950196-2-6

health impacts include mortality and morbidity from extreme heat events, injuries and displacement from floods and storms, and respiratory harm from wildfire smoke. Indirect impacts include the geographic expansion of vector-borne diseases such as malaria, dengue fever, and Lyme disease as warming temperatures allow disease vectors to colonize higher altitudes and latitudes; the disruption of agricultural systems and consequent food insecurity; and the mental health consequences of displacement, loss, and climate-related trauma.

Watts et al. (2015)[5] argued that climate change threatens to undermine the public health gains of the past half century, reversing progress on child mortality, infectious disease control, and food security in the most vulnerable regions of the world. They called for a comprehensive health response to climate change that treats it not only as an environmental problem but as a health emergency requiring the engagement of the global medical and public health community.

Conclusion

The evidence is unambiguous: human health and environmental health are inseparable. From the air that people breathe to the water they drink, from the ecological systems that regulate disease vectors to the biodiversity that provides pharmaceutical resources, the conditions of the natural world continuously shape the conditions of human well-being. Effective medicine must engage with ecology, and effective environmental policy must recognize human health as one of its central objectives. The One Health framework provides a promising conceptual foundation for this integration, but realizing its potential requires institutional changes in health systems, research funding, and policy processes that remain substantial challenges for the coming decades.

References

1. [1] McMichael, A. J. (2017). *Climate change and the health of nations.* Oxford University Press.
2. [2] Carpenter, S. R., et al. (1998). Nonpoint pollution of surface waters with phosphorus and nitrogen. *Ecological Applications*, 568–559 ,(3)*8*.
3. [3] Zinsstag, J., et al. (2011). From "one medicine" to "one health" and systemic approaches to health and well-being. *Preventive Veterinary Medicine,* –148 ,(4–3)*101* 156.
4. [4] Preston, C. J. (2011). *The synthetic age: Outdesigning evolution, resurrecting species, and reengineering our world.* MIT Press.
5. [5] Watts, N., et al. (2015). Health and climate change: Policy responses to protect public health. *The Lancet,* 1914–1861 ,(10006)*386*.

Isbn: 979-8-9950196-2-6

Chapter 7

Traditional Ethnobotanical Knowledge: From Local Practices to Global Perspectives on Conservation and Sustainability

Dr. Pitambar Humane

Professor and Head,
Department of Botany
Dharampeth M. P. Deo Memorial Science College,
Nagpur-440033
Email Id - aareenpapa_2004@rediffmail.com

Abstract

Ethnobotany — the study of the relationships between human societies and plants — occupies a unique position at the boundary of natural science, social science, and humanities. Traditional ethnobotanical knowledge, accumulated by indigenous and local communities over millennia of intimate observation and interaction with plant life, represents a priceless repository of information about plant diversity, ecological relationships, medicinal

Isbn: 979-8-9950196-2-6

properties, and sustainable use practices. This chapter traces the development of ethnobotany as a scientific discipline, examines the depth and diversity of traditional plant knowledge systems, explores the pharmacological and ecological significance of ethnobotanical knowledge for modern science and conservation, and addresses the critical ethical and legal questions surrounding the appropriation, protection, and benefit-sharing of indigenous plant knowledge. The chapter argues that the conservation of traditional ethnobotanical knowledge is inseparable from the conservation of the plant diversity it describes and the cultural diversity of the communities that hold it.

Keywords: Ethnobotany, traditional knowledge, indigenous communities, medicinal plants, bioprospecting, Convention on Biological Diversity, conservation, benefit-sharing

Introduction

Plants are foundational to human life — as sources of food, medicine, fiber, fuel, shelter, and spiritual meaning. The knowledge that human communities have developed about the plants in their environments over thousands of years of cohabitation constitutes one of the richest and most practically significant bodies of traditional knowledge that exists. Ethnobotany, as a formal scientific discipline, emerged in the late nineteenth century with the work of scholars such as Edward Palmer and Matilda Coxe Stevenson, who documented the plant knowledge of indigenous communities in North America. Since then, it has grown into a vibrant interdisciplinary field with deep connections to botany, pharmacology, ecology, anthropology, and conservation biology.

Isbn: 979-8-9950196-2-6

Traditional ethnobotanical knowledge is not static — it is a dynamic, evolving system of understanding, continuously refined through observation, experimentation, and cultural transmission. Communities that have inhabited particular landscapes for generations develop highly detailed knowledge of local plant diversity, ecological relationships, seasonal patterns, and sustainable harvest practices that ecological scientists are only beginning to recognize and document. Moerman (1998)[1] documented nearly 30,000 uses of more than 2,500 plant species by indigenous peoples of North America alone — a body of knowledge representing an extraordinary investment of human observation and understanding accumulated over countless generations.

7.1 The Scope and Depth of Traditional Plant Knowledge

Traditional ethnobotanical knowledge encompasses far more than a catalogue of useful plants. It includes detailed ecological knowledge of plant communities and their dynamics, sophisticated understanding of plant-animal interactions, knowledge of the environmental conditions associated with particular plant species, and understanding of the relationships between plant use, harvest intensity, and long-term plant population viability.

Key dimensions of traditional plant knowledge include:

- **Medicinal plant knowledge:** Identification, preparation, and therapeutic application of plants for a vast range of ailments, including conditions for which modern medicine has no or limited treatments

Isbn: 979-8-9950196-2-6

- **Agricultural plant knowledge:** Cultivation techniques, seed selection and saving, intercropping practices, and soil management developed over centuries of farming experience
- **Ecological plant knowledge:** Understanding of plant succession, indicator species, seasonal phenology, and the relationships between plant diversity and ecosystem function
- **Cultural and spiritual plant knowledge:** The roles of plants in ritual, ceremony, identity, and cosmology — knowledge systems that embed plant use within broader cultural frameworks of meaning and responsibility

Balick and Cox (1996)[2] estimated that fewer than one percent of flowering plant species have been thoroughly investigated for their pharmacological potential, and that traditional knowledge systems provide the most efficient guide to identifying plant species most likely to yield bioactive compounds of therapeutic value.

7.2 Ethnobotany and Pharmacology

The pharmacological significance of traditional plant knowledge is perhaps its most widely recognized dimension in the contemporary scientific world. Approximately half of all modern pharmaceutical drugs were derived originally from natural compounds, and a substantial proportion of these were identified through leads provided by traditional medicine systems.

The antimalarial compound artemisinin, derived from *Artemisia annua* (sweet wormwood), is one of the most celebrated examples. Knowledge of this plant's fever-reducing properties was documented in Chinese traditional medical texts dating to the fourth century CE, and it was this traditional knowledge that guided the Nobel Prize-winning research of Tu Youyou in identifying artemisinin as the active compound and developing it into the world's most effective antimalarial treatment. Similarly, the cardiac drug digoxin was derived from *Digitalis purpurea* (foxglove), long used in European folk medicine for heart conditions.

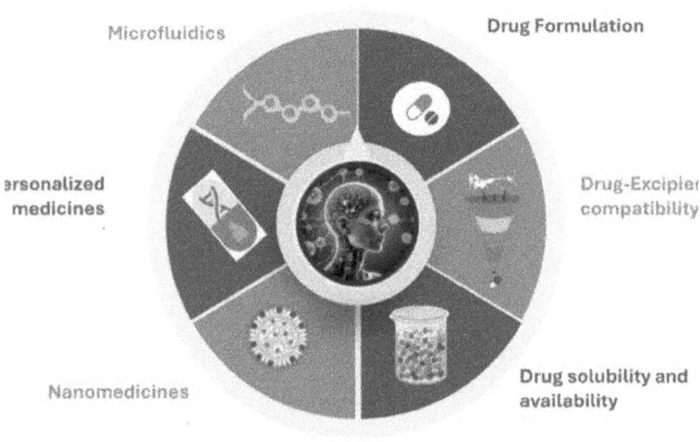

Figure 7: Pathway from Traditional Plant Knowledge to Modern Pharmaceutical Development:

7.3 Conservation Implications of Ethnobotanical Knowledge

The relationship between ethnobotanical knowledge and plant conservation is bidirectional and mutually reinforcing. Traditional communities that maintain detailed knowledge of plant diversity are typically also active stewards of the habitats in which that diversity exists. Sacred forests, traditional agricultural

Isbn: 979-8-9950196-2-6

landscapes, and customary resource management systems maintained by indigenous communities have been shown to harbor significant plant diversity, often exceeding that found in formally protected areas in the same regions.

Conversely, the erosion of traditional knowledge — driven by cultural assimilation, language loss, displacement, and the marginalization of traditional practices — typically accompanies and accelerates the loss of the plant diversity it describes and sustains. Maffi (2005)[3] documented the striking parallel between global maps of linguistic diversity and maps of biological diversity, arguing that biocultural diversity — the diversity of life in all its manifestations, biological and cultural — is a single phenomenon requiring integrated conservation responses.

Conservation strategies that engage indigenous and local communities as active participants and knowledge holders — rather than treating them as problems to be excluded from protected areas — consistently show better outcomes for both biodiversity and community well-being. Tuxill and Nabhan (2001)[4] document numerous case studies where community-based conservation approaches combining traditional knowledge with scientific monitoring have achieved significant successes in maintaining plant diversity in agricultural and wild landscapes.

7.4 Ethics, Intellectual Property, and Benefit-Sharing

The global recognition of traditional ethnobotanical knowledge as both scientifically valuable and commercially significant has raised profound ethical and legal questions about ownership, access, and the equitable sharing of benefits arising from its use.

Isbn: 979-8-9950196-2-6

The appropriation of traditional plant knowledge by pharmaceutical and biotechnology companies without consent or compensation to the communities that developed and maintain that knowledge — a practice commonly described as **biopiracy** — has been extensively documented and vigorously contested.

The Convention on Biological Diversity (CBD) and its Nagoya Protocol on Access and Benefit-Sharing, adopted in 2010, establish an international framework requiring that access to genetic resources and associated traditional knowledge be subject to prior informed consent from the communities holding that knowledge, and that benefits arising from commercial use be shared equitably. Dutfield (2004)[5] provides a comprehensive analysis of the intellectual property dimensions of this framework, examining its strengths, limitations, and implementation challenges.

Conclusion

Traditional ethnobotanical knowledge is a living heritage of incalculable value — to science, to medicine, to conservation, and to the cultural identities of the communities that hold and transmit it. Its documentation, protection, and ethical engagement by the scientific community are matters of both intellectual and moral urgency. The plants and the knowledge about them are in many cases disappearing at comparable rates, driven by the same forces of cultural erosion, habitat destruction, and economic marginalization. Effective conservation of plant diversity must therefore be inseparable from the protection and revitalization of the cultural systems within which that knowledge lives and is sustained.

Isbn: 979-8-9950196-2-6

References

1. [1] Moerman, D. E. (1998). *Native American ethnobotany.* Timber Press.
2. [2] Balick, M. J., & Cox, P. A. (1996). *Plants, people, and culture: The science of ethnobotany.* Scientific American Library.
3. [3] Maffi, L. (2005). Linguistic, cultural, and biological diversity. *Annual Review of Anthropology,* 617–599 *,34.*
4. [4] Tuxill, J., & Nabhan, G. P. (2001). *People, plants, and protected areas.* Earthscan.
5. [5] Dutfield, G. (2004). *Intellectual property, biogenetic resources, and traditional knowledge.* Earthscan.

Isbn: 979-8-9950196-2-6

Chapter 8

Eco-Management: Business, Commerce, and Environmental Responsibility

Prof. Mahesh Bajirao Hajare

Assistant Professor, Sahakar Maharshi Bhausaheb Santuji Thorat Arts, Science and Commerce College, Sangamner. Dist.-Ahilyanagar
Email Id - maheshhajare24296@gmail.com

Abstract

The relationship between business enterprise and the natural environment has undergone a fundamental transformation in recent decades. Once largely ignored or treated as an external constraint, environmental responsibility has moved to the center of corporate strategy, investment analysis, and regulatory compliance. This chapter examines the emergence and evolution of eco-management as both a conceptual framework and a practical discipline, tracing the development of corporate environmental responsibility from early compliance-based approaches to the sophisticated integration of environmental performance into business strategy, supply chain management, and financial reporting. It discusses key frameworks including Environmental Management Systems, Corporate Social Responsibility, the circular economy, Environmental, Social, and Governance (ESG) investing, and green supply chains, and evaluates the evidence on whether voluntary corporate environmental action can achieve the scale of transformation that ecological sustainability requires.

Keywords: Eco-management, corporate environmental responsibility, environmental management systems, ESG, green economy, circular economy, sustainable business

Introduction

Businesses are among the most powerful actors shaping the environmental conditions of the planet. Through their decisions about what to produce, how to produce it, where to source materials, how to manage waste, and how to engage with regulatory systems, corporations exercise enormous influence over patterns of resource consumption, pollution, land use, and greenhouse gas emission. For most of industrial history, this influence was exercised with little attention to environmental consequences — the assumption that natural resources were inexhaustible inputs and the natural environment an infinite sink for waste was built into the operating logic of industrial capitalism.

The emergence of eco-management — the systematic integration of environmental considerations into business management — represents a significant departure from this tradition. Driven by a combination of regulatory pressure, consumer demand, investor scrutiny, and a growing recognition among business leaders that long-term profitability depends on ecological stability, eco-management has evolved from a marginal concern of

Isbn: 979-8-9950196-2-6

specialist environmental departments into a strategic priority engaging the highest levels of corporate leadership. Esty and Winston (2006)[1] documented this transformation in their account of how leading companies were using environmental strategy to drive innovation, reduce costs, and build competitive advantage.

8.1 Environmental Management Systems

Environmental Management Systems (EMS) provide a structured framework for organizations to systematically manage their environmental impacts, continuously improve their environmental performance, and demonstrate compliance with applicable environmental regulations. The ISO 14001 standard, first published by the International Organization for Standardization in 1996, is the most widely adopted EMS framework globally, with hundreds of thousands of organizations certified to the standard across more than 170 countries.

The ISO 14001 framework is built around a Plan-Do-Check-Act cycle:

- **Plan:** Identify environmental aspects and impacts of activities, set environmental objectives and targets, and develop plans for achieving them
- **Do:** Implement the environmental management plan, including training, communication, and operational controls
- **Check:** Monitor and measure environmental performance against objectives and targets, conduct internal audits

Isbn: 979-8-9950196-2-6

- **Act:** Take corrective action, review the system at senior management level, and use review outputs to drive continuous improvement

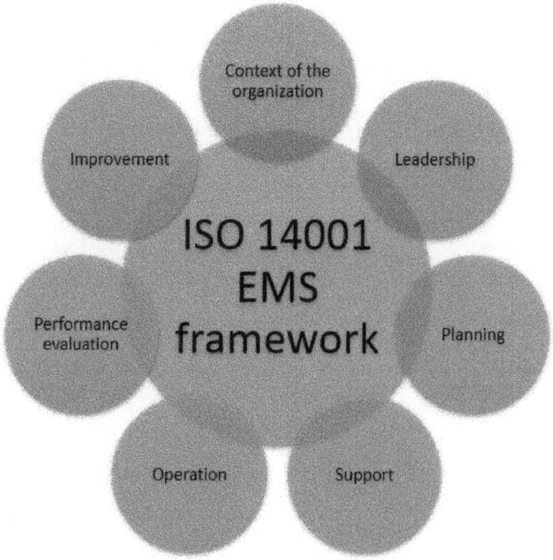

Figure 8: ISO 14001 Environmental Management System Cycle:

8.2 Corporate Environmental Responsibility and ESG

Corporate Social Responsibility (CSR) frameworks have progressively incorporated environmental performance alongside social and governance dimensions, evolving into the Environmental, Social, and Governance (ESG) paradigm that now dominates the language of sustainable business and responsible investment. ESG metrics provide investors, analysts, and other stakeholders with standardized information about corporate environmental performance — including greenhouse gas emissions,

Isbn: 979-8-9950196-2-6

water use, waste generation, and supply chain environmental risks — alongside social and governance indicators.

The growth of ESG investing has been dramatic. Global sustainable investment now represents tens of trillions of dollars in assets under management, and major institutional investors — including sovereign wealth funds and pension funds — have committed to integrating ESG criteria into their investment decision-making. Friede, Busch, and Bassen (2015)[2] conducted a comprehensive meta-analysis of the relationship between ESG performance and financial performance, finding a positive relationship in the majority of studies — suggesting that environmental responsibility and financial performance are complementary rather than contradictory objectives.

8.3 Green Supply Chain Management

Global supply chains are among the most significant channels through which business activity drives environmental impact. The sourcing of raw materials, the manufacturing of components, the transportation of goods, and the management of end-of-life products all involve environmental impacts that extend far beyond the direct operations of any individual company. Green Supply Chain Management (GSCM) involves the integration of environmental criteria into supply chain decisions across all of these stages.

Key GSCM practices include:

- Supplier environmental assessment and auditing
- Green procurement criteria that reward environmentally responsible suppliers

- Design for environment principles that reduce material consumption and improve recyclability
- Reverse logistics systems for product take-back, reuse, and recycling
- Transportation optimization to reduce fuel consumption and emissions

Srivastava (2007)[3] provided an early comprehensive review of GSCM, tracing its development from a concern focused primarily on regulatory compliance toward a strategic approach linking environmental performance with supply chain efficiency and innovation.

8.4 The Business Case for Sustainability

The question of whether environmental responsibility creates or destroys business value has been extensively debated and researched. The evidence increasingly supports the view that, beyond a certain threshold of regulatory and reputational risk management, genuine environmental leadership creates competitive advantages through cost reduction, innovation stimulation, talent attraction, and brand differentiation.

Porter and van der Linde (1995)[4] made the influential argument that environmental regulation, when well designed, stimulates innovation that ultimately enhances rather than undermines competitiveness — the so-called Porter Hypothesis. Companies that are forced to find cleaner ways of producing, or that voluntarily adopt more resource-efficient processes, often discover that the resulting innovations reduce costs, improve quality, and open new markets.

Isbn: 979-8-9950196-2-6

The circular economy provides perhaps the most compelling business case for environmental sustainability, framing waste reduction not as a cost but as an efficiency gain — an opportunity to extract more value from the same material inputs through closed-loop design, reuse, and recycling. Ellen MacArthur Foundation (2013)[5] estimated that the transition to a circular economy in Europe could represent net material cost savings of hundreds of billions of euros annually, alongside significant reductions in greenhouse gas emissions and resource consumption.

Conclusion

Eco-management has matured from a peripheral concern into a central dimension of business strategy, driven by regulatory pressure, investor expectations, consumer demand, and a growing recognition that long-term business success depends on environmental stability. The tools available for corporate environmental management — from ISO 14001 to ESG reporting to green supply chain standards — have become increasingly sophisticated. The remaining challenge is one of scale and ambition: whether voluntary corporate action, even at its most progressive, can achieve the fundamental transformation of production and consumption systems that ecological sustainability requires, or whether regulatory frameworks must be substantially strengthened to ensure that environmental costs are systematically internalized across the economy.

References

1. [1] Esty, D., & Winston, A. (2006). *Green to gold: How smart companies use environmental strategy to innovate,*

Isbn: 979-8-9950196-2-6

create value, and build competitive advantage. Yale University Press.

2. [2] Friede, G., Busch, T., & Bassen, A. (2015). ESG and financial performance: Aggregated evidence from more than 2000 empirical studies. *Journal of Sustainable Finance & Investment,* 233–210 ,(4)*5*.
3. [3] Srivastava, S. K. (2007). Green supply-chain management: A state-of-the-art literature review. *International Journal of Management Reviews,* –53 ,(1)*9* 80.
4. [4] Porter, M. E., & van der Linde, C. (1995). Toward a new conception of the environment-competitiveness relationship. *Journal of Economic Perspectives,* –97 ,(4)*9* 118.
5. [5] Ellen MacArthur Foundation. (2013). *Towards the circular economy: Economic and business rationale for an accelerated transition.* Ellen MacArthur Foundation.

Isbn: 979-8-9950196-2-6

Chapter 9

Eco-Technology: Integrating Computer Science, AI, and Green Innovations

Dr. Vinod Kumar Singh (Associate professor)

K.G.K (PG) college moradabad pin-code (244001)

Email Id - vinod.ap81@yahoo.in.

Abstract

The intersection of digital technology and environmental sustainability is producing a new paradigm of eco-technology — the application of computational tools, artificial intelligence, sensor networks, and digital platforms to the monitoring, modeling, and management of environmental systems. This chapter examines how computer science and artificial intelligence are being deployed in service of environmental goals, from real-time pollution monitoring and biodiversity tracking to climate modeling, smart grid optimization, and precision agriculture. It also addresses the ecological paradox at the heart of digital technology — that the infrastructure of the digital economy itself consumes enormous quantities of energy and materials — and examines efforts to develop greener computing and more sustainable digital

Isbn: 979-8-9950196-2-6

systems. The chapter argues that technology is neither inherently green nor inherently destructive, but that its environmental trajectory depends on the values, priorities, and governance structures within which it is developed and deployed.

Keywords: Eco-technology, artificial intelligence, environmental monitoring, climate modeling, smart systems, green computing, precision agriculture, digital sustainability

Introduction

Technology has always shaped the human relationship with the natural environment, from the invention of the plow to the industrial steam engine to the nuclear reactor. The digital revolution of the late twentieth and early twenty-first centuries has added an entirely new dimension to this relationship, creating tools of unprecedented power for understanding, monitoring, modeling, and managing environmental systems — while simultaneously generating new forms of environmental pressure through the energy and material demands of digital infrastructure.

The concept of eco-technology encompasses both the application of digital tools to environmental challenges and the effort to make digital technology itself more environmentally sustainable. Hilty and Aebischer (2015)[1] provide a framework for thinking about the relationship between information and communication technology and sustainability, distinguishing between first-order effects (the direct environmental impacts of producing and operating ICT equipment), second-order effects (the environmental impacts of activities enabled or displaced by ICT use), and third-order effects (the systemic changes in economic and social structures driven by ICT over the long term).

9.1 Environmental Monitoring and Remote Sensing

One of the most transformative contributions of digital technology to environmental management has been the development of remote sensing and environmental monitoring systems capable of observing the Earth's surface, atmosphere, and oceans at spatial and temporal scales impossible with ground-based methods alone.

Satellite-based remote sensing now provides continuous, global coverage of vegetation health, land cover change, sea surface temperature, ice extent, atmospheric composition, and a wide range of other environmental variables. The Landsat program, operating continuously since 1972, has produced the longest continuous record of Earth's land surface from space, enabling scientists to document half a century of land cover change, urban expansion, deforestation, glacier retreat, and vegetation dynamics.

Ground-based sensor networks — including air quality monitoring stations, river flow gauges, ocean buoys, and soil moisture sensors — complement satellite observations by providing high-temporal-resolution data at specific locations. The Internet of Things (IoT) is rapidly expanding the density and diversity of environmental sensor networks, enabling near-real-time monitoring of environmental conditions at scales from individual buildings to entire river catchments.

Isbn: 979-8-9950196-2-6

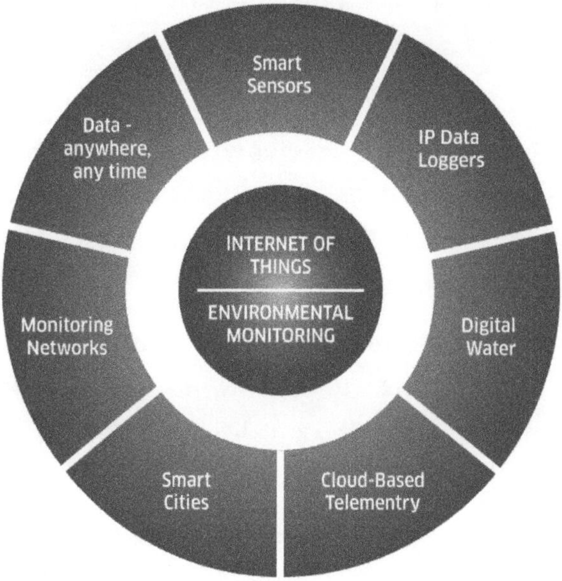

Figure 9: Environmental Monitoring Technology Architecture:

9.2 Artificial Intelligence in Environmental Applications

Artificial intelligence and machine learning are being applied to an expanding range of environmental challenges, exploiting their capacity to identify patterns in large, complex datasets that exceed human analytical capacity.

Key applications include:

- **Species identification and biodiversity monitoring:** Computer vision algorithms trained on image datasets can identify plant and animal species from photographs, acoustic recordings, or video with accuracy comparable to or exceeding human experts, enabling

Isbn: 979-8-9950196-2-6

citizen science programs to contribute effectively to large-scale biodiversity monitoring

- **Climate downscaling:** Machine learning models can translate coarse-resolution global climate model outputs into high-resolution local projections relevant for agricultural planning, urban infrastructure design, and ecosystem management
- **Deforestation detection:** Convolutional neural networks applied to satellite imagery can detect forest cover change in near real-time, enabling rapid response to illegal clearing events
- **Energy system optimization:** AI algorithms optimize the operation of renewable energy systems, balancing supply variability from solar and wind with demand fluctuations to maximize clean energy utilization
- **Precision agriculture:** Machine learning applied to remote sensing, soil sensor, and weather data enables site-specific management of irrigation, fertilization, and pest control, reducing input use and environmental impact while maintaining or improving yields

Rolnick et al. (2019)[2] provided a comprehensive survey of machine learning applications in climate change mitigation and adaptation, identifying dozens of high-potential application areas across energy, transportation, buildings, agriculture, and carbon management.

Isbn: 979-8-9950196-2-6

9.3 Green Computing and Digital Sustainability

The environmental paradox of digital technology is that the infrastructure enabling eco-technological applications is itself a significant and growing source of environmental impact. Data centers consumed approximately 200 terawatt-hours of electricity globally in 2018 — roughly one percent of global electricity demand — and this figure is growing rapidly with the expansion of cloud computing, video streaming, and AI computation. The manufacturing of electronic devices requires significant quantities of rare earth elements, precious metals, and other materials whose extraction involves serious environmental and social impacts.

Masanet et al. (2020)[3] documented the decoupling of data center energy consumption from computing workloads achieved through efficiency improvements in server hardware, cooling systems, and software virtualization, but cautioned that continued efficiency gains may be insufficient to offset the growth in computational demand driven by AI and cloud services.

Green computing initiatives aim to reduce the environmental footprint of digital infrastructure through renewable energy procurement, energy efficiency improvements, hardware lifespan extension, and responsible end-of-life management of electronic equipment. Major technology companies including Google, Microsoft, and Amazon have made significant renewable energy commitments for their data center operations, though the full lifecycle environmental impacts of digital infrastructure remain a subject of ongoing research and debate.

Isbn: 979-8-9950196-2-6

9.4 Smart Cities and Green Infrastructure

The application of digital technology to urban systems — the smart city concept — offers significant opportunities for reducing the environmental footprint of cities, which account for approximately 70 percent of global greenhouse gas emissions while housing more than half the world's population. Smart city applications relevant to environmental sustainability include:

- Smart grid systems that optimize electricity distribution, integrate distributed renewable energy sources, and enable demand response
- Intelligent transportation systems that reduce traffic congestion, optimize public transit, and support the integration of electric vehicles
- Smart building energy management systems that optimize heating, cooling, and lighting in response to occupancy and weather conditions
- Urban metabolism monitoring systems that track material and energy flows through cities to identify efficiency improvement opportunities

Bibri and Krogstie (2017)[4] examined the relationship between smart city technologies and urban sustainability, finding significant potential for environmental improvement but also highlighting risks of increased energy consumption from digital infrastructure and the exclusion of lower-income communities from smart city benefits.

Isbn: 979-8-9950196-2-6

Conclusion

Eco-technology represents both a genuine opportunity and a genuine challenge for environmental sustainability. The computational tools now available for monitoring, modeling, and managing environmental systems are extraordinary in their power and reach — enabling scientists to understand ecological dynamics at scales and resolutions previously unimaginable, and enabling managers and policymakers to make better-informed decisions about resource allocation and environmental protection. At the same time, the digital infrastructure underlying these capabilities carries its own significant environmental footprint, which must be systematically addressed through green computing strategies, renewable energy procurement, and circular approaches to hardware production and management. The environmental value of eco-technology ultimately depends on whether its benefits for sustainability are larger than its costs — a question that requires careful empirical assessment rather than technological optimism.

References

1. [1] Hilty, L. M., & Aebischer, B. (2015). *ICT innovations for sustainability.* Springer.
2. [2] Rolnick, D., et al. (2019). Tackling climate change with machine learning. *ACM Computing Surveys,* ,(2)*53* 96–1.
3. [3] Masanet, E., et al. (2020). Recalibrating global data center energy-use estimates. *Science,* –984 ,(6481)*367* 986.
4. [4] Bibri, S. E., & Krogstie, J. (2017). Smart sustainable cities of the future: An extensive interdisciplinary

Isbn: 979-8-9950196-2-6

literature review. *Sustainable Cities and Society*, –183 ,*31* 212.

Isbn: 979-8-9950196-2-6

Chapter 10

Environment and Humanity: An Interdisciplinary Approach

Dr. Akhilesh Sharma

Dr. Akhilesh Sharma
Associate professor
Shivpati P. G. College
Shohratgarh, Siddharthnagar
Pin code. 272205
Mobile no. 9984127177
Email Id - akhilesh8831@gmail.com

Abstract

The relationship between humanity and the natural environment is the most fundamental and consequential relationship in the history of life on Earth. This chapter takes an explicitly interdisciplinary perspective on this relationship, examining how different academic traditions — from evolutionary biology and cultural anthropology to philosophy, history, and sociology — have understood the human place in nature and the ways in which human societies have shaped, been shaped by, and are now profoundly threatening the natural systems that sustain them. It traces the deep history of human-environment relationships from prehistoric hunter-gatherer societies through agricultural

Isbn: 979-8-9950196-2-6

civilizations, industrialization, and the contemporary era of global ecological change, and examines the cultural, philosophical, and social dimensions that determine whether humanity can develop the collective will and institutional capacity for a sustainable future.

Keywords: Human ecology, cultural history of environment, anthropocene, human-nature relationships, environmental sociology, sustainability transitions, deep history

Introduction

Humanity is a biological species — shaped by evolution, dependent on ecological systems, and subject to the same fundamental constraints of energy, matter, and biological function that govern all life on Earth. At the same time, humanity is unique among all species in the scale and intentionality of its environmental influence, in its capacity to reflect on that influence, and in its potential to consciously choose different relationships with the natural world. This dual nature — ecological creature and reflective agent — defines the central paradox of the human relationship with the environment.

The concept of the **Anthropocene** — a proposed geological epoch defined by the dominant influence of human activity on Earth's geology, atmosphere, and biosphere — captures the scale of this transformation. Crutzen and Stoermer (2000)[1] introduced this term to reflect the scientific consensus that human activities have become the primary driver of global environmental change, altering atmospheric chemistry, ocean chemistry, land cover, species composition, and even the geological record in ways that will be detectable in the stratigraphic record for millions of years.

Isbn: 979-8-9950196-2-6

10.1 Deep History: Human-Environment Relationships Across Time

Understanding the contemporary environmental crisis requires a long view — one that places the current moment within the deep history of human-environment relationships stretching back hundreds of thousands of years. Archaeological and paleontological evidence documents a long history of human impacts on ecosystems, from the megafaunal extinctions associated with the spread of anatomically modern humans out of Africa beginning approximately 70,000 years ago, through the landscape transformations associated with the adoption of agriculture beginning around 10,000 years BCE, to the dramatically accelerating impacts of industrialization over the past two centuries.

Ponting (2007)[2] provides a sweeping history of human-environment relationships from the earliest hunter-gatherer societies to the present, documenting a consistent pattern in which human communities have exploited the ecological resources of their environments to the limits of technological capacity, often overshooting those limits and triggering environmental degradation and social crisis. He argues that this pattern, repeated across cultures and centuries, reflects deep structural features of human social organization and economic motivation rather than the ignorance or irrationality of particular historical actors.

The transition to agriculture — arguably the most consequential transformation in human-environment relationships in prehistory — fundamentally altered the ecological basis of human life, enabling larger, denser populations while simultaneously creating new forms of environmental vulnerability through dependence on monoculture crops, exposure to animal-borne diseases,

Isbn: 979-8-9950196-2-6

and the progressive degradation of soil fertility through cultivation.

10.2 Culture, Values, and Environmental Behavior

Human behavior toward the natural environment is shaped not only by economic incentives and institutional constraints but by deeply rooted cultural values, worldviews, and cosmologies that determine how people understand their relationship to the natural world. These cultural dimensions of human-environment relationships have been studied extensively by anthropologists, sociologists, and environmental humanists, who have documented the extraordinary diversity of ways in which different cultures have conceptualized nature, defined appropriate relationships with it, and structured their resource use practices accordingly.

White (1967)[3] made the provocative argument that the Judeo-Christian tradition, with its emphasis on human dominion over nature and its desacralization of the natural world, bore significant historical responsibility for Western ecological attitudes that facilitated exploitation without restraint. While this argument has been extensively debated and refined, it opened an important conversation about the role of worldviews and cultural values in shaping environmental behavior — a conversation that extends to the examination of indigenous ecological worldviews, Eastern philosophical traditions, and contemporary secular environmental ethics.

Isbn: 979-8-9950196-2-6

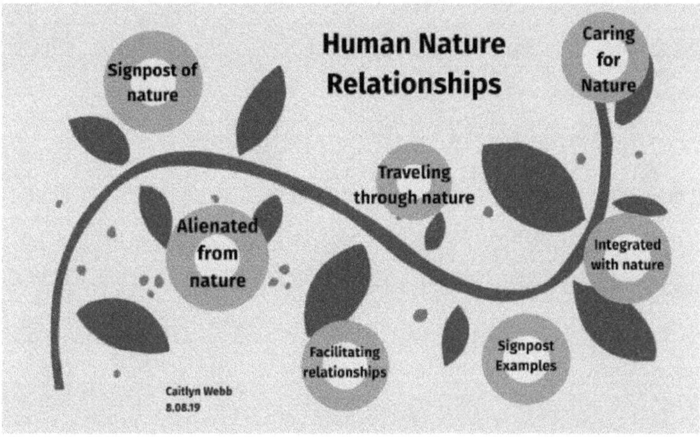

Figure 10: Cultural Dimensions of Human-Nature Relationships:

10.3 Environmental Sociology and Social Drivers of Environmental Change

Environmental sociology examines the social structures, institutions, and processes that drive environmental change and mediate human responses to it. The Treadmill of Production theory, developed by Schnaiberg (1980)[4], argues that the structural dynamics of capitalist economies — the imperative of profit accumulation, the demands of workers for wages, and the requirements of the state for tax revenue — create a self-reinforcing cycle of increasing production and consumption that continuously expands the withdrawal of resources from ecological systems and the addition of wastes to them, regardless of the environmental intentions of individual actors within the system.

This structural analysis has significant implications for sustainability transitions. If environmental degradation is driven by structural features of economic organization rather than by individual attitudes or behaviors, then policy responses focused

Isbn: 979-8-9950196-2-6

primarily on changing individual behavior — through education, information campaigns, and voluntary action — are insufficient without parallel structural changes in economic institutions, incentive systems, and governance frameworks.

Ecological modernization theory, developed by Mol and Spaargaren (2000)[5] among others, offers a more optimistic structural account, arguing that capitalist industrial systems are capable of self-reform toward greater ecological rationality through technological innovation, institutional change, and the progressive internalization of environmental values into economic decision-making. The empirical debate between these theoretical frameworks — whether existing economic structures can be reformed toward sustainability or whether more fundamental transformation is required — remains one of the central unresolved questions in environmental social science.

10.4 Toward a Sustainable Human Future

The aspiration for a sustainable human future — one in which human well-being is maintained without degrading the ecological systems that sustain it — requires more than technological innovation or policy reform. It requires a cultural transformation in how societies understand success, progress, and the good life, and institutional transformation in how economic systems distribute resources, allocate costs, and govern the commons.

The concept of **planetary stewardship** — the idea that humanity has both the capacity and the responsibility to consciously manage its collective relationship with the Earth system — has been advanced by scholars including Folke et al. (2011) as a response to the realities of the Anthropocene. It acknowledges

Isbn: 979-8-9950196-2-6

that there is no longer any pristine nature untouched by human influence, and that the task is therefore not to return to some imagined pre-human ecological state but to navigate the Anthropocene in ways that maintain the stability and resilience of Earth systems within which human civilization can flourish.

Conclusion

The relationship between environment and humanity is the most consequential question of our time. Understanding it requires the full depth of human knowledge — evolutionary biology, history, anthropology, sociology, philosophy, economics, and the natural sciences — brought to bear on a reality of extraordinary complexity. The Anthropocene confronts humanity with a challenge unlike any it has faced before: not the challenge of surviving in a natural world that is indifferent to human welfare, but the challenge of governing a human-altered world in ways that maintain the ecological foundations of civilization. Meeting this challenge requires not only better science and better technology but a genuine transformation in the cultural values and institutional structures through which human societies relate to the natural world.

References

1. [1] Crutzen, P. J., & Stoermer, E. F. (2000). The Anthropocene. *Global Change Newsletter*, 18–17 ,*41*.
2. [2] Ponting, C. (2007). *A new green history of the world: The environment and the collapse of great civilizations.* Penguin.

Isbn: 979-8-9950196-2-6

3. [3] White, L. (1967). The historical roots of our ecological crisis. *Science,* 1207–1203 ,(3767)*155.*
4. [4] Schnaiberg, A. (1980). *The environment: From surplus to scarcity.* Oxford University Press.
5. [5] Mol, A. P. J., & Spaargaren, G. (2000). Ecological modernisation theory in debate: A review. *Environmental Politics,* 49–17 ,(1)*9.*

Isbn: 979-8-9950196-2-6

Chapter 11

Eco-Integration: Linking Technology, Humanities, and Life Sciences

Shayni Devendra Nikam

Shayni Devendra Nikam
Assistant professor
K.A.A.N.M.S.Arts, Commerce, Science College Satana
Pin- 423301
snikam706@gmail.com

Abstract

Eco-integration refers to the deliberate and systematic effort to bring together the conceptual frameworks, methods, and insights of technology, the humanities, and the life sciences in service of a more complete and actionable understanding of environmental sustainability. This chapter examines the intellectual and practical case for eco-integration, exploring how the combination of quantitative scientific data, humanistic interpretation, and technological application produces insights that none of these domains can generate in isolation. It traces examples of successful eco-integration across domains including conservation biology and indigenous knowledge, environmental data science and

Isbn: 979-8-9950196-2-6

cultural history, and green engineering and ecological ethics, arguing that the fragmentation of knowledge across disciplinary silos is itself a barrier to sustainability and that overcoming this fragmentation is both an intellectual and institutional imperative.

Keywords: Eco-integration, interdisciplinary ecology, technology and humanities, biocultural diversity, systems thinking, sustainability science, knowledge integration

Introduction

The natural environment does not organize itself according to the disciplinary boundaries of the modern university. Ecosystems integrate physical, chemical, biological, and social processes in ways that no single academic discipline is equipped to fully describe or explain. Yet the production of knowledge about environments — and the training of professionals who manage them — remains overwhelmingly organized around disciplinary specialization, rewarding depth within fields over the integrative thinking that environmental challenges demand.

Eco-integration challenges this organizational logic by asserting that meaningful environmental understanding and effective environmental action require the deliberate combination of multiple knowledge traditions. It is not simply interdisciplinarity in the sense of combining two or more disciplines, but a more fundamental rethinking of how knowledge is organized, produced, and applied in relation to environmental systems. Wilson (1998)[1] anticipated this need in his concept of **consilience** — the unity of knowledge across disciplines achieved through the discovery of deep structural relationships between them — though eco-integration goes further in emphasizing the practical and policy

Isbn: 979-8-9950196-2-6

dimensions of knowledge integration alongside its intellectual dimensions.

11.1 Technology Meets Life Sciences: Environmental Data Science

The life sciences — ecology, evolutionary biology, physiology, genetics, and microbiology — generate the fundamental knowledge about living systems that underpins environmental understanding. Technology, particularly digital computing, remote sensing, and genomics, has transformed the scale and resolution at which this knowledge can be gathered and analyzed, creating new opportunities for discovery while also generating new challenges of data integration and interpretation.

Environmental genomics — the application of genomic sequencing technologies to the study of organisms and communities in their natural environments — illustrates this convergence powerfully. Metagenomic approaches allow scientists to characterize the full diversity of microbial communities in soils, water, and air from environmental samples without culturing, revealing ecological diversity of extraordinary richness that was invisible to conventional microbiological methods. This technology-enabled revolution in microbial ecology has transformed understanding of nutrient cycling, soil health, water quality, and the ecological functions of microbial communities.

Key areas where technology and life sciences converge in environmental research include:

- Remote sensing and landscape ecology — satellite and drone-based data integrated with field ecological sur-

Isbn: 979-8-9950196-2-6

veys to understand landscape-scale patterns of biodiversity and ecosystem function

- Genomics and conservation biology — population genomic approaches identifying connectivity, inbreeding risk, and adaptive variation in threatened species populations
- Environmental sensor networks and phenology — automated monitoring of plant flowering, animal migration, and seasonal ecological events in response to climate change
- Bioinformatics and ecosystem modeling — computational analysis of large biological datasets to build predictive models of ecosystem response to environmental change

11.2 Humanities and Life Sciences: Biocultural Integration

The integration of humanistic and life science perspectives on environment is less commonly discussed than the technology-science interface, but it is equally important for developing a complete understanding of human-environment relationships. The humanities — including history, literature, philosophy, cultural studies, and the arts — contribute dimensions of understanding that life science alone cannot provide: the historical depth of human-environment interactions, the cultural diversity of environmental values and practices, the narrative and aesthetic dimensions of human engagement with nature, and the ethical frameworks for evaluating environmental choices.

Isbn: 979-8-9950196-2-6

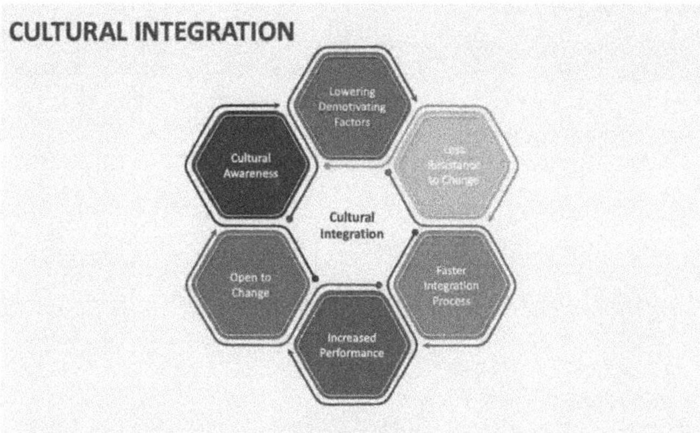

Figure 11: Biocultural Diversity Integration Model:

The **environmental humanities** — a rapidly growing interdisciplinary field bringing together history, literature, philosophy, and cultural studies — provide critical perspectives on how human societies have understood, valued, and managed their environments across time and cultures. Rose et al. (2012)[2] described the environmental humanities as a response to the recognition that environmental problems are as much cultural and ethical as they are scientific and technical, and that addressing them requires the interpretive and critical resources of humanistic scholarship alongside the empirical resources of natural science.

11.3 Technology and Humanities: Digital Environmental Communication

The communication of environmental knowledge — translating scientific findings into forms that can inform public understanding, political deliberation, and cultural change — is one of the most significant practical challenges in the sustainability field.

Isbn: 979-8-9950196-2-6

Here, the integration of technology and humanities offers particularly promising possibilities.

Digital storytelling, data visualization, interactive mapping, and immersive media experiences can make complex environmental data accessible and emotionally resonant to audiences far beyond the specialist scientific community. Climate change, abstract in its global scale and long time horizons, can be communicated through local, personal, and narrative-rich forms that connect with lived experience in ways that statistics alone cannot achieve. Pahl-Wostl et al. (2013)[3] examined the role of knowledge exchange and communication in sustainability transitions, finding that the form and framing of environmental knowledge — not just its content — significantly affects its capacity to influence decision-making.

Environmental citizen science platforms exemplify this convergence of technology and humanistic engagement, mobilizing public participation in data collection while simultaneously building environmental literacy, connection to nature, and civic engagement with conservation issues. Apps like iNaturalist, eBird, and Globe Observer have enabled millions of non-specialist participants to contribute observations of biodiversity, weather, and land cover that complement and expand professional scientific monitoring.

11.4 Systems Thinking as Eco-Integration Framework

Systems thinking provides perhaps the most developed intellectual framework for eco-integration — one that moves beyond the combination of disciplinary perspectives toward a genuinely

Isbn: 979-8-9950196-2-6

integrative understanding of complex adaptive systems that are simultaneously ecological, social, technological, and cultural. The concept of **social-ecological systems**, developed most fully by Ostrom and colleagues, treats human communities and their ecological environments as coupled adaptive systems in which feedbacks operate in both directions and in which understanding or managing one side of the system in isolation from the other consistently produces unexpected and often counterproductive outcomes.

Meadows (2008)[4] articulated the core principles of systems thinking in an accessible form applicable across environmental contexts, emphasizing the importance of understanding system structure — particularly feedback loops, delays, and nonlinearities — as the key to anticipating and influencing system behavior. Her framework for identifying high-leverage points in systems — the places where small interventions can produce large changes — provides a practical tool for eco-integration in policy and management contexts.

Folke et al. (2010)[5] developed the concept of **resilience thinking** as a framework for managing social-ecological systems, emphasizing the importance of maintaining the diversity, modularity, and adaptive capacity of these systems in the face of disturbance and change. Resilience thinking integrates ecological, social, and governance dimensions of sustainability in a framework that has been widely applied in fisheries management, forest governance, water resource management, and urban planning.

Isbn: 979-8-9950196-2-6

Conclusion

Eco-integration is not an academic luxury but an intellectual and practical necessity. The environmental challenges of the Anthropocene — complex, multi-scaled, deeply uncertain, and deeply value-laden — cannot be understood or addressed through any single disciplinary lens. The deliberate integration of technology, humanities, and life sciences in service of environmental understanding and sustainability practice represents one of the most important intellectual developments of the contemporary era. Its institutionalization — in universities, research programs, policy processes, and professional training — requires sustained investment and cultural change within the knowledge institutions that shape how societies understand and respond to their environmental condition.

References

1. [1] Wilson, E. O. (1998). *Consilience: The unity of knowledge.* Knopf.
2. [2] Rose, D. B., et al. (2012). Thinking through the environment, unsettling the humanities. *Environmental Humanities,* 5–1 ,(1)*1.*
3. [3] Pahl-Wostl, C., et al. (2013). Environmental flows and water governance: Managing flows. *Ecology and Society,* 26 ,(4)*18.*
4. [4] Meadows, D. H. (2008). *Thinking in systems: A primer.* Chelsea Green Publishing.

Isbn: 979-8-9950196-2-6

5. [5] Folke, C., et al. (2010). Resilience thinking: Integrating resilience, adaptability, and transformability. *Ecology and Society,* 20 ,(4)*15*.

Isbn: 979-8-9950196-2-6

Chapter 12

Household Chemistry in Traditional Indian Life

Dr. Pratibha Gopalrao Raundal

Pratibha Gopalrao Raundal
Assistant professor
K.A.A.N.M.S.Arts, Commerce, Science College Satana
Pin- 423301
Email Id - pratibhabhadane14@gmail.com

Abstract

Traditional Indian households have, for millennia, employed a sophisticated repertoire of chemical substances and processes in daily life — from cooking and food preservation to cleaning, personal hygiene, textile care, ritual practice, and medicine. These practices, embedded in cultural knowledge systems transmitted across generations, represent a form of applied chemistry that predates formal scientific inquiry and often anticipates principles that modern chemistry has subsequently confirmed. This chapter explores the chemical knowledge embedded in traditional Indian household practices, examining the substances used, the processes involved, and the ecological and health wisdom that guided their selection and application. It further considers how the

Isbn: 979-8-9950196-2-6

replacement of traditional household chemistry by synthetic chemical products over the course of the twentieth century has altered both the ecological footprint and the cultural knowledge landscape of Indian domestic life, and what the recovery and revaluation of traditional practices might contribute to contemporary sustainability agendas.

Keywords: Traditional knowledge, household chemistry, India, Ayurveda, natural cleaning agents, food preservation, fermentation, plant-based materials, sustainability

Introduction

The Indian household has historically been a site of remarkable chemical knowledge and practice. Long before the emergence of industrial chemistry, Indian families managed complex chemical processes daily — fermenting foods, extracting dyes, making soaps, preserving medicines, purifying water, and maintaining health through carefully formulated preparations drawn from plant, mineral, and animal sources. This knowledge was not abstract or theoretical but deeply practical, embedded in the rhythms of daily life and transmitted from generation to generation through practice, observation, and oral instruction.

The Ayurvedic tradition provides the most systematically documented framework for understanding the chemical dimensions of traditional Indian life, offering a sophisticated account of the properties of natural substances and their interactions with the human body and the environment. However, the chemical knowledge of traditional Indian households extends well beyond the Ayurvedic medical tradition to encompass a vast body of domestic practice — what might be called the everyday chemistry

Isbn: 979-8-9950196-2-6

of traditional life — that has received far less scholarly attention but is equally significant as a repository of ecological and chemical wisdom. Valiathan (2003)[1] provides a scholarly account of the Ayurvedic tradition's engagement with natural substances, documenting the depth and sophistication of traditional pharmacological knowledge.

12.1 Plant-Based Cleaning Agents and Domestic Chemistry

The cleaning agents of traditional Indian households were derived almost entirely from plant sources, each chosen for specific chemical properties that modern chemistry can now explain. Several of the most widely used traditional cleaning substances demonstrate this chemical sophistication clearly.

Soapnut (Reetha — *Sapindus mukorossi*): The pericarp of the soapnut fruit contains saponins — naturally occurring surfactant molecules with a hydrophilic (water-attracting) sugar component and a hydrophobic (fat-attracting) aglycone component. This amphiphilic molecular structure allows saponins to function exactly as synthetic surfactants do in modern detergents, lowering surface tension, emulsifying oils and fats, and suspending dirt particles in water for rinsing away. Soapnuts were used across India for washing hair, clothing, and utensils, demonstrating that traditional practice had independently identified natural surfactant chemistry without any formal molecular understanding.

Shikakai (*Acacia concinna*): Used primarily as a hair cleanser, shikakai pods contain saponins, vitamin C, and various flavonoids. Its mildly acidic pH is gentle on hair and scalp, and its saponin content provides cleansing action while the vitamins and

Isbn: 979-8-9950196-2-6

antioxidants condition the hair fiber. Modern shampoo formulations attempting to replicate the benefits of shikakai have confirmed the chemical rationale behind its traditional use.

Ash and Lime (*Chuna*): Wood ash, rich in potassium carbonate (K_2CO_3) and calcium oxide (CaO), was widely used in traditional Indian households for cleaning utensils, purifying water, and treating leather. When dissolved in water, these alkaline compounds raise the pH of the cleaning solution, saponifying fats and dissolving organic grime through a chemical mechanism identical to that exploited by modern alkaline cleaning products. Lime (*chuna* — calcium hydroxide, $Ca(OH)_2$) was similarly used for whitewashing walls (providing antimicrobial surfaces through its high alkalinity), treating grains, and preparing betel (paan).

Key traditional cleaning agents and their active chemistry:

- Neem leaves (*Azadirachta indica*) — azadirachtin and nimbidin compounds providing antimicrobial and insecticidal activity
- Turmeric (*Curcuma longa*) — curcumin providing anti-inflammatory, antimicrobial, and antioxidant activity; also used as a natural dye
- Tamarind (*Tamarindus indica*) — tartaric acid providing mild acid cleaning action for metal polishing
- Besan (chickpea flour) — proteins and mild exfoliants used in skin care preparations

Isbn: 979-8-9950196-2-6

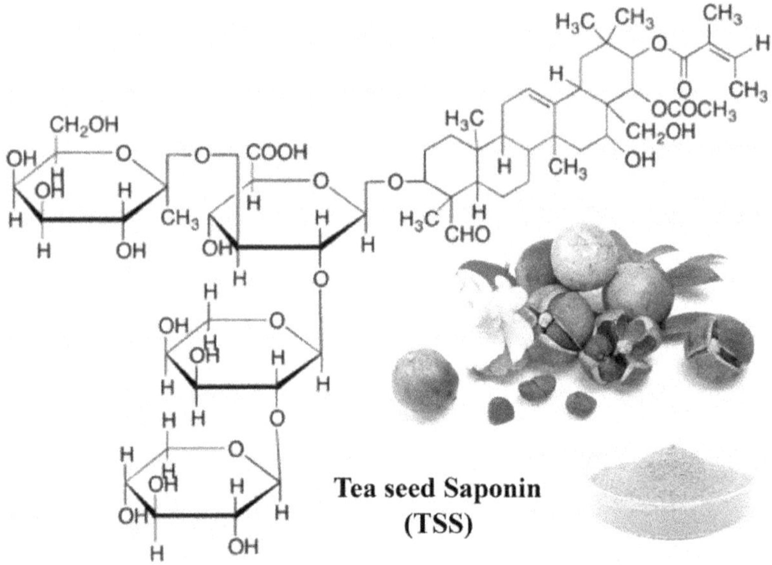

Figure 12: Chemical Structure of Saponin Molecule as Used in Soapnut:

12.2 Food Preservation and Fermentation Chemistry

Traditional Indian food preservation represents one of the richest domains of applied household chemistry. Before refrigeration, Indian households developed sophisticated methods for extending the shelf life of foods through fermentation, acidification, drying, salting, oiling, and the use of antimicrobial spices — each exploiting specific chemical principles to inhibit microbial spoilage.

Fermentation: Fermented foods are central to Indian cuisine across regions — idli and dosa (fermented rice and lentil batters), dhokla (fermented chickpea batter), kanji (fermented carrot

Isbn: 979-8-9950196-2-6

drink), various pickles and chutneys fermented with lactic acid bacteria, and fermented dairy products including curd (dahi) and butter milk (chaas). The fermentation process involves the metabolic activity of bacteria and yeasts that convert carbohydrates into organic acids and alcohols, lowering the pH of the food and creating an environment hostile to pathogenic microorganisms. Lactic acid fermentation also enhances nutritional value — increasing bioavailability of minerals, producing B vitamins, and generating probiotic organisms with beneficial effects on gut health.

Pickling with Oil, Salt, and Spices: The Indian pickle (achar) tradition exploits multiple chemical preservation mechanisms simultaneously. Salt draws water out of food tissues through osmosis, lowering water activity and inhibiting microbial growth. Oil creates an anaerobic environment hostile to aerobic spoilage organisms. Spices including turmeric, mustard, fenugreek, and asafoetida contribute antimicrobial compounds — including curcumin, allyl isothiocyanate, and various phenolic compounds — that further suppress microbial activity. The combination of these mechanisms produces preserved foods that can remain stable for months or years at ambient temperature.

12.3 Traditional Water Purification

Access to clean water is a perennial challenge, and Indian traditions developed several effective methods for improving water quality in domestic settings. Copper vessels were widely used for water storage — a practice whose antimicrobial rationale has been confirmed by modern research demonstrating that copper surfaces release cupric ions (Cu^{2+}) that are toxic to bacteria and other

Isbn: 979-8-9950196-2-6

microorganisms at very low concentrations. The practice of storing water in clay pots exploited the cooling effect of evaporation through the porous clay surface, as well as the mild adsorptive properties of clay minerals in removing certain contaminants.

The use of **alum (phitkari — potassium aluminum sulfate, $KAl(SO_4)_2 \cdot 12H_2O$)** for water clarification represents a particularly clear example of traditional chemical knowledge anticipating modern water treatment. When alum is dissolved in water, the aluminum ions hydrolyze to form aluminum hydroxide ($Al(OH)_3$), which precipitates as a gelatinous floc that aggregates fine suspended particles and carries them to the bottom of the vessel — the same principle of coagulation-flocculation that underlies modern municipal water treatment. Muddy or turbid water treated with alum clarifies dramatically, and the process also removes many pathogenic organisms carried on suspended particles.

12.4 Ritual and Medicinal Household Chemistry

The ritual dimensions of traditional Indian household chemistry are inseparable from its practical and medicinal dimensions. The burning of camphor (*Cinnamomum camphora*) in religious rituals releases volatile organic compounds with antimicrobial properties, fumigating the domestic environment; the smearing of doorways and floors with turmeric paste provides antimicrobial protection at the threshold of the home; the application of sindoor (*vermilion*) and other mineral-based cosmetics reflects knowledge of the optical and adhesive properties of mineral pigments developed over centuries of use.

Isbn: 979-8-9950196-2-6

Ayurvedic household medicine drew on an extensive pharmacopoeia of plant-based preparations for managing common ailments. The use of ginger (*Zingiber officinale*) for digestive complaints, tulsi (*Ocimum tenuiflorum*) for respiratory infections, neem (*Azadirachta indica*) for skin conditions, and ashwagandha (*Withania somnifera*) for stress and fatigue all reflect accumulated empirical knowledge of the pharmacological activity of these plants — knowledge that modern phytochemistry has progressively confirmed and elaborated. Patwardhan (2005)[2] provides a systematic account of the convergences between Ayurvedic traditional medicine and modern pharmacology, documenting the scientific basis of many traditional preparations.

Conclusion

Traditional Indian household chemistry represents a sophisticated, ecologically grounded body of applied knowledge developed over millennia of careful observation and cultural transmission. Its cleaning agents, food preservation practices, water purification methods, and medicinal preparations demonstrate a deep practical understanding of chemical principles that modern science can now explain at the molecular level. The progressive replacement of these traditional practices by synthetic chemical products over the course of the twentieth century has brought convenience but has also displaced ecological knowledge, increased chemical exposure, and generated waste streams incompatible with environmental sustainability. The recovery, revaluation, and thoughtful modern application of traditional household chemical knowledge — informed by both cultural wisdom and scientific understanding — represents an important dimension of the broader project of sustainable living in the twenty-first century.

Isbn: 979-8-9950196-2-6

References

1. [1] Valiathan, M. S. (2003). *The legacy of Charaka.* Orient Longman.
2. [2] Patwardhan, B. (2005). Ethnopharmacology and drug discovery. *Journal of Ethnopharmacology,* –50 ,(2–1)*100* 52.
3. [3] Samal, P. K., et al. (2010). Local health traditions of Uttarakhand. *Indian Journal of Traditional Knowledge,* 338–333 ,(2)*9.*
4. [4] Kumar, V., & Chopra, A. K. (2012). Phytochemical analysis and antimicrobial potential of *Acacia concinna. Journal of Applied and Natural Science,* 106–101 ,(1)*4.*
5. [5] Aggarwal, B. B., & Harikumar, K. B. (2009). Potential therapeutic effects of curcumin. *International Journal of Biochemistry and Cell Biology,* 59–40 ,(1)*41.*

Isbn: 979-8-9950196-2-6

Chapter 13

Molecular Mechanisms in Zoological Diversity

Smt. Vaishali Ashok Rahane

Assistant professor
College - GMD College sinnar
Pin- 422104
Email Id- vaishalirahane222@gmail.com

Abstract

Zoological diversity, encompassing the stunning array of animal body plans, life histories, and behaviors, is fundamentally an emergent property of molecular processes. For centuries, zoology focused on macro-level descriptions — classification, morphology, and ecology. However, the last few decades have seen a revolution, catalyzed by genomics, that bridges the gap between the structure of a DNA sequence and the structure of an elephant's trunk or a jellyfish's stinging cell. This chapter explores the pivotal molecular mechanisms that generate, maintain, and modify the astonishing variety of animal life, moving beyond simply cataloging diversity to understanding its fundamental, underlying instruction set. We will examine how gene regulation,

Isbn: 979-8-9950196-2-6

genomic architecture, and molecular pathways drive evolution and adaptation across the animal kingdom.

Keywords: Gene regulatory networks, Hox genes, cis-regulatory elements, genomic duplication, transposable elements, epigenetics, molecular evolution, zoological diversity

Introduction

Zoological diversity, encompassing the stunning array of animal body plans, life histories, and behaviors, is fundamentally an emergent property of molecular processes. For centuries, zoology focused on macro-level descriptions — classification, morphology, and ecology. However, the last few decades have seen a revolution, catalyzed by genomics, that bridges the gap between the structure of a DNA sequence and the structure of an elephant's trunk or a jellyfish's stinging cell. This chapter explores the pivotal **molecular mechanisms** that generate, maintain, and modify the astonishing variety of animal life, moving beyond simply cataloging diversity to understanding its fundamental, underlying instruction set. We will examine how gene regulation, genomic architecture, and molecular pathways drive evolution and adaptation across the animal kingdom.

The Role of Gene Regulation in Morphological Novelty

The primary driver of macro-level zoological diversity is not the sudden appearance of entirely new genes, but rather the **re-engineering of existing gene regulatory networks (GRNs)**. The classic example lies in the *Hox* genes, a conserved set of transcription factors responsible for specifying body axis identity

Isbn: 979-8-9950196-2-6

across nearly all Bilaterians. While a mouse and a fly share homologous *Hox* clusters, the vast morphological differences between them — from segmented appendages to complex vertebral columns — arise largely from *where* and *when* these genes are expressed, controlled by **Cis-Regulatory Elements (CREs)**.

CREs, such as enhancers and promoters, act as molecular switches, integrating signals to modulate the transcription rate of their target genes. Minor changes in the sequence or position of a CRE can dramatically alter developmental patterns, leading to significant evolutionary changes, or *morphological novelty*. For instance:

- The loss of limbs in snakes involved mutations in CREs controlling the expression of the *Sonic Hedgehog (Shh)* gene in the developing limb bud.
- The evolution of the arthropod carapace or the different beak shapes in Darwin's finches can be traced to modifications in the timing and location of master regulatory gene expression, often mediated by subtle changes in distant CREs.

This principle emphasizes that the evolution of new body forms is frequently a story of *regulatory evolution* rather than *protein-coding evolution*. The animal genome is less a rigid blueprint and more an exquisitely tunable circuit board, allowing for significant phenotypical plasticity and evolutionary experimentation. Carroll (2005)[1] elaborated this principle extensively in his account of the role of developmental gene regulation in the evolution of animal form.

Isbn: 979-8-9950196-2-6

Genomic Architecture and Speciation

The large-scale organization of the animal genome itself provides molecular mechanisms for both stabilizing species identity and facilitating evolutionary change, thereby driving diversity. Key mechanisms include:

Horizontal Gene Transfer (HGT)

While more common in single-celled organisms, HGT plays a sporadic but powerful role in zoological innovation. Examples include the transfer of genes from bacteria or fungi to arthropods, providing novel metabolic pathways or toxin resistance genes.

Whole Genome Duplication (WGD) and Gene Duplication

The duplication of entire genomes (WGD) or individual genes is a major source of raw genetic material. WGD, notably observed in the early evolution of vertebrates (the "two rounds of WGD" hypothesis), provided a redundancy that allowed one copy of the duplicated genes to retain the original function while the other copy was free to acquire a new function (**neofunctionalization**) or specialize one aspect of the original function (**subfunctionalization**). This mechanism is fundamental to the evolution of complex systems, such as the vertebrate immune system or sensory perception arrays. Ohno (1970)[2] provided the foundational theoretical account of gene duplication as the primary mechanism of evolutionary innovation.

Isbn: 979-8-9950196-2-6

Transposable Elements (TEs)

Often dismissed as "junk DNA," TEs are mobile genetic elements that can jump around the genome. TEs can act as evolutionary catalysts in several ways:

1. **Insertion:** They can insert into or near genes, disrupting or changing expression patterns.
2. **Gene Capture:** They can carry adjacent genomic fragments, contributing to gene duplication.
3. **CRE Generation:** Crucially, TEs can contribute novel CREs, providing entirely new regulatory inputs for existing genes. For example, specific TE insertions have been linked to the evolution of placental development in mammals.

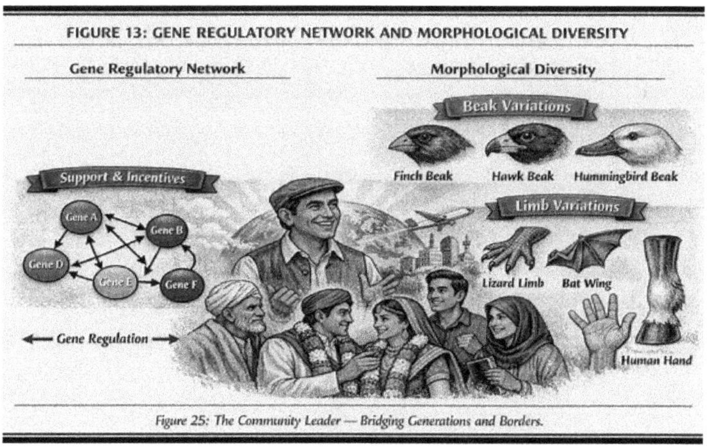

Figure 13: Gene Regulatory Network and Morphological Diversity:

Isbn: 979-8-9950196-2-6

Molecular Mechanisms of Adaptation and Environmental Response

Zoological diversity is also a reflection of an animal's capacity to adapt to myriad environments, from deep-sea vents to high-altitude plateaus. Molecular mechanisms facilitate this immediate and long-term adaptation.

Molecular Physiology and Gene Expression Plasticity

Molecular physiology explores how genes and proteins mediate organismal function. **Acclimation** to short-term environmental changes (e.g., temperature, salinity) is achieved by rapidly adjusting the expression of stress-response genes. For instance, animals exposed to extreme heat activate **Heat Shock Proteins (HSPs)**, molecular chaperones that prevent irreversible damage to cellular proteins. The ability to fine-tune the proteome through rapid transcriptional and translational control underpins physiological adaptability and contributes to niche specialization. Feder and Hofmann (1999)[3] provide a comprehensive account of HSP function and its role in thermal tolerance across animal taxa.

Epigenetics and Transgenerational Inheritance

Epigenetic mechanisms — changes in gene function that do not involve changes in the DNA sequence — provide a flexible link between the environment and the genome. These include **DNA methylation** and **histone modification**, which can silence or activate specific genes. Epigenetic marks are often responsive

Isbn: 979-8-9950196-2-6

to environmental cues (e.g., diet, stress, temperature) and can, in some cases, be inherited across generations. This **transgenerational epigenetic inheritance** is a mechanism for rapidly adjusting offspring phenotypes to a predictable change in the environment, bypassing the slow process of classic natural selection driven by DNA mutations. Jablonka and Lamb (2005)[4] developed a comprehensive account of epigenetic inheritance and its evolutionary implications, arguing that epigenetic variation constitutes a second, non-genetic system of inheritance with significant consequences for evolutionary theory.

Conclusion

The study of zoological diversity has moved irrevocably into the molecular realm. The four core mechanisms discussed — **re-engineering of GRNs**, **genomic structural innovation (duplication/HGT)**, **physiological plasticity (HSPs)**, and **epigenetic inheritance** — provide the ultimate explanation for the differences between a sponge, a shark, and a sparrow. Understanding the complexity of animal life requires a synthesis of molecular data with traditional zoological observations. The future of zoology lies in **Systems Biology**, where high-throughput sequencing data (genomics, transcriptomics, proteomics) are integrated with ecological and behavioral studies. By identifying the specific CREs that control adaptive traits, the molecular pathways that enable extreme physiological feats, and the epigenetic marks that prime generations for environmental challenge, we can gain a profound and predictive understanding of how animal diversity arose and how it can be conserved in the face of accelerating global change.

Isbn: 979-8-9950196-2-6

References

1. [1] Carroll, S. B. (2005). *Endless forms most beautiful: The new science of evo devo.* W. W. Norton.
2. [2] Ohno, S. (1970). *Evolution by gene duplication.* Springer.
3. [3] Feder, M. E., & Hofmann, G. E. (1999). Heat-shock proteins, molecular chaperones, and the stress response. *Annual Review of Physiology,* 282–243 ,*61.*
4. [4] Jablonka, E., & Lamb, M. J. (2005). *Evolution in four dimensions: Genetic, epigenetic, behavioral, and symbolic variation in the history of life.* MIT Press.
5. [5] Carroll, S. B., Grenier, J. K., & Weatherbee, S. D. (2005). *From DNA to diversity: Molecular genetics and the evolution of animal design* (2nd ed.). Blackwell.

Isbn: 979-8-9950196-2-6

Chapter 14

Digital Sustainability: The Role of Data, Cloud, and Cyber Sciences in Environment

Dr. Syed Shahid Raza

Syed Shahid Raza,
Assistant Professor,
CMS B-School, JAIN(Deemed-to-be University), pin code : 560009
Email Id- dr.syed_shahidraza@cms.ac.in

Abstract

The digital revolution has transformed virtually every dimension of human activity, and its relationship with environmental sustainability is complex, dual-natured, and rapidly evolving. On one hand, digital technologies — including big data analytics, cloud computing, the Internet of Things, and cybersecurity infrastructure — offer powerful tools for monitoring environmental systems, optimizing resource use, enabling clean energy transitions, and supporting evidence-based environmental governance. On the other hand, the infrastructure of the digital economy is

Isbn: 979-8-9950196-2-6

itself a significant and growing consumer of energy and material resources, raising important questions about the net environmental impact of digital transformation. This chapter examines both dimensions of this relationship, exploring how data, cloud computing, and cyber sciences can be harnessed in service of environmental sustainability while simultaneously addressing the ecological footprint of digital infrastructure itself. It argues for a framework of digital sustainability that treats environmental performance as a core design criterion for digital systems from the outset.

Keywords: Digital sustainability, cloud computing, big data, Internet of Things, cyber-physical systems, green data centers, environmental monitoring, digital carbon footprint

Introduction

The world generates data at an extraordinary and accelerating rate. Every sensor, satellite, smartphone, connected vehicle, industrial machine, and social media interaction adds to a global data stream that is already measured in zettabytes and is projected to grow by orders of magnitude over the coming decade. This data, when appropriately collected, managed, and analyzed, represents an unprecedented resource for understanding and managing environmental systems — from tracking deforestation in real time to optimizing the dispatch of renewable energy generators to predicting flood events days in advance.

At the same time, the infrastructure required to store, process, and transmit this data — millions of servers in data centers consuming vast quantities of electricity, billions of electronic devices manufactured from rare and often ecologically harmful materials,

Isbn: 979-8-9950196-2-6

and global telecommunications networks running continuously — represents a significant environmental footprint in its own right. Andrae and Edler (2015)[1] projected that the information and communication technology sector could account for a substantial share of global electricity consumption by 2030 if current growth trajectories continue without efficiency improvements.

The concept of **digital sustainability** — which this chapter takes as its organizing framework — addresses both dimensions of this relationship, asking how digital technologies can be designed, deployed, and governed to maximize their environmental benefits while minimizing their environmental costs.

14.1 Big Data and Environmental Monitoring

The capacity to collect, store, and analyze large volumes of environmental data has transformed the scope and resolution of environmental science and management. Satellite remote sensing programs now generate petabytes of imagery annually covering the Earth's land surface, oceans, and atmosphere. Ground-based sensor networks contribute additional streams of high-temporal-resolution data on air quality, water quality, soil moisture, biodiversity indicators, and energy system performance. Social media and citizen science platforms add qualitative and semi-quantitative data on environmental conditions and events.

The value of this data for environmental management depends critically on the analytical methods available for extracting meaningful signals from high-dimensional, noisy, and often incomplete datasets. Machine learning methods — including neural

Isbn: 979-8-9950196-2-6

networks, random forests, and clustering algorithms — have proven particularly powerful for environmental data analysis, enabling:

- Detection of deforestation events from satellite imagery with high accuracy and minimal latency
- Prediction of air quality conditions from combinations of sensor data, meteorological variables, and land use information
- Classification of marine habitats from acoustic and optical data collected by autonomous underwater vehicles
- Early warning of coral bleaching events from sea surface temperature anomalies and water quality data

Reichstein et al. (2019)[2] examined the application of deep learning to Earth system science, documenting both the extraordinary potential and the important limitations of machine learning approaches in environmental contexts — particularly their dependence on large training datasets and their relative opacity compared to process-based models.

14.2 Cloud Computing and Environmental Applications

Cloud computing — the delivery of computing services including servers, storage, databases, networking, and software over the internet — has democratized access to large-scale computational resources for environmental research and management. Research teams and organizations that previously lacked the

Isbn: 979-8-9950196-2-6

infrastructure to process large environmental datasets can now access computing resources on demand, paying only for what they use and scaling up or down according to their needs.

Key environmental applications of cloud computing include:

- **Climate modeling at scale:** Global climate models require enormous computational resources. Cloud platforms enable research institutions and national meteorological agencies to run high-resolution climate projections that would previously have required dedicated supercomputing facilities.
- **Disaster response and resilience:** Cloud-based platforms integrate real-time satellite, sensor, and social media data to support rapid response to natural disasters including floods, wildfires, and cyclones, improving the speed and effectiveness of emergency management.
- **Environmental data sharing and interoperability:** Cloud platforms facilitate the sharing of environmental datasets across institutional and national boundaries, enabling collaborative research at global scales.

The environmental footprint of cloud computing is significant but is declining relative to the computational work performed, due to the large efficiency gains achieved by hyperscale data center operators compared to the distributed on-premise computing infrastructure that cloud services have replaced. Masanet et al. (2020)[3] documented this efficiency advantage, finding that cloud services can perform the same computational tasks with significantly less energy than equivalent on-premise

Isbn: 979-8-9950196-2-6

infrastructure — though this advantage is partially or wholly offset by the growth in total computational demand.

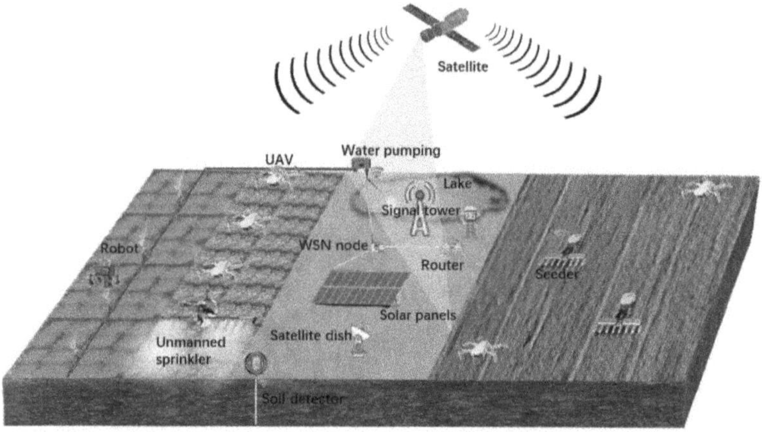

Figure 14: Cloud-Based Environmental Data Pipeline:

14.3 Cybersecurity and Environmental Infrastructure

The increasing dependence of environmental and energy infrastructure on digital systems creates significant cybersecurity risks that have direct environmental implications. Smart grids, water treatment systems, industrial pollution control equipment, and environmental monitoring networks are all increasingly networked and therefore potentially vulnerable to cyber attack. Successful attacks on these systems could disrupt energy supply, contaminate water supplies, disable pollution monitoring, or create industrial accidents with serious environmental consequences.

Cybersecurity for environmental infrastructure therefore represents an important and underappreciated dimension of digital

Isbn: 979-8-9950196-2-6

sustainability. Protecting the integrity, availability, and confidentiality of environmental data and control systems is not merely a technical concern but a sustainability imperative. The field of **cyber-physical systems security** addresses the specific challenges of protecting systems in which digital networks directly control physical infrastructure, developing methods for detecting intrusions, maintaining system resilience, and ensuring safe operation even when digital components are compromised.

At the same time, cybersecurity practices themselves have environmental dimensions. Cryptographic operations, which are computationally intensive, contribute significantly to the energy consumption of networked systems. The energy consumption of blockchain networks — which use cryptographic proof-of-work mechanisms — has attracted particular attention, with some estimates suggesting that the Bitcoin network alone consumes more electricity annually than many medium-sized countries. De Vries (2018)[4] provided an early analysis of Bitcoin's energy consumption and its environmental implications, stimulating broader debate about the energy efficiency of distributed ledger technologies.

14.4 Green Data Centers and Sustainable Digital Infrastructure

The environmental footprint of digital infrastructure is concentrated primarily in data centers, where the electricity consumption of servers, cooling systems, and power delivery infrastructure is substantial and growing. The Power Usage Effectiveness (PUE) metric, which compares total data center energy consumption to the energy consumed by IT equipment

Isbn: 979-8-9950196-2-6

alone, has improved significantly over the past decade as operators have invested in more efficient cooling systems, server hardware, and power delivery infrastructure.

Beyond energy efficiency, sustainable data center design addresses:

- **Renewable energy procurement:** Leading data center operators have made significant commitments to sourcing electricity from renewable energy sources, either through direct investment in renewable energy projects or through power purchase agreements with renewable generators.
- **Water use efficiency:** Data center cooling systems are significant consumers of freshwater, and water use efficiency is increasingly recognized as an environmental priority alongside energy efficiency, particularly in water-stressed regions.
- **Circular economy principles in hardware management:** Extending the useful life of server hardware, refurbishing equipment for secondary use, and ensuring responsible recycling of end-of-life equipment are increasingly recognized as important dimensions of data center environmental management.
- **Location decisions:** Siting data centers in regions with cool climates (reducing cooling energy requirements), access to renewable energy, and adequate water resources is an important lever for reducing the environmental footprint of digital infrastructure.

Isbn: 979-8-9950196-2-6

Whitehead, Andrews, and Shah (2015)[5] examined the lifecycle environmental impacts of cloud computing services, finding that the significant efficiency advantages of cloud over on-premise computing are real but context-dependent, and that the continued growth in demand for cloud services requires ongoing investment in efficiency and renewable energy to prevent total environmental impact from growing.

Conclusion

Digital sustainability is not simply about using technology to monitor and manage the environment — it is about ensuring that the digital systems themselves are developed and operated in ways that are compatible with environmental sustainability. The extraordinary potential of big data, cloud computing, and cyber sciences for environmental monitoring, clean energy optimization, and evidence-based governance is real and important. But this potential can only be fully realized if the ecological footprint of digital infrastructure is systematically addressed through renewable energy, efficiency improvement, circular economy approaches to hardware management, and thoughtful governance of the total resource demands of a digitizing global economy. Digital sustainability, understood in this comprehensive sense, is one of the defining challenges of the Anthropocene — and one of its most promising opportunities.

References

1. [1] Andrae, A. S. G., & Edler, T. (2015). On global electricity usage of communication technology: Trends to 2030. *Challenges*, 157–117 ,(1)*6*.

Isbn: 979-8-9950196-2-6

2. [2] Reichstein, M., et al. (2019). Deep learning and process understanding for data-driven Earth system science. *Nature*, 204–195 ,*566*.
3. [3] Masanet, E., et al. (2020). Recalibrating global data center energy-use estimates. *Science*, –984 ,(6481)*367* 986.
4. [4] De Vries, A. (2018). Bitcoin's growing energy problem. *Joule*, 805–801 ,(5)*2*.
5. [5] Whitehead, B., Andrews, D., & Shah, A. (2015). Assessing the environmental impact of data centers. *Journal of Industrial Ecology*, 221–211 ,(2)*19*.

Isbn: 979-8-9950196-2-6

Chapter 15

The Algorithmic Zoologist: How Artificial Intelligence and Machine Learning are Redefining Zoological Research.

Dr.Trupti Dhananjay Kakulte

Assistant professor
K.A.A.N.M.S.Arts, Commerce, Science College Satana
Pin- 423301
Email Id-saanvitrupti@gmail.com

The discipline of zoology, centred on the magnificent and diverse life of the animal kingdom, is presently experiencing a paradigm shift. Traditionally reliant on painstaking manual observation, taxonomy, and field work, the field is now being rapidly transformed by the integration of Artificial Intelligence (AI) and Machine Learning (ML). These powerful computational tools are not merely assisting zoologists; they are fundamentally changing how data is collected, analysed, and translated into scientific insight and effective conservation action. This chapter explores

Isbn: 979-8-9950196-2-6

how AI and ML are becoming indispensable instruments across the spectrum of zoological research.

1. The Digital Sentinel: AI in Wildlife Monitoring and Conservation

The most immediate and impactful application of AI is in addressing the global biodiversity crisis by enhancing wildlife monitoring and conservation. The need for automated systems stems from the sheer volume of data generated by modern field techniques.

Automated Census and Surveillance-

- **Computer Vision for Species Identification:** Camera traps and drones deployed in remote ecosystems generate millions of images and videos. Manually sorting these files is an insurmountable task. Deep Learning models, particularly **Convolutional Neural Networks (CNNs)**, are trained to process these visuals, automatically identifying species, counting individuals, and even recognizing unique individuals based on patterns like stripes (zebras), spots (leopards), or natural scars. This enables real-time population estimates and trend analysis with unprecedented speed and objectivity.
- **Bioacoustics for Elusive Animals:** Many species are nocturnal or inhabit dense environments, making visual monitoring impossible. AI models analyze continuous audio recordings (bioacoustics) to detect, classify, and localize animal vocalizations, from the songs of

Isbn: 979-8-9950196-2-6

migratory birds to the clicks of deep-sea cetaceans. This allows for the non-invasive study of species presence, density, and communication patterns over vast spatial and temporal scales.

Predictive Conservation Strategies

Beyond simple counting, ML models empower proactive conservation. By analyzing historical poaching incidents, terrain data, and weather patterns, **predictive policing** algorithms can forecast high-risk areas. This strategic intelligence allows conservation teams to optimize patrol routes and resource allocation, making anti-poaching efforts vastly more efficient and targeted.

2. Unlocking the Black Box: AI in Animal Behavior and Ethology

Studying animal behavior (**ethology**) has long required detailed, often subjective, observation. AI/ML now offers tools to quantify complex behaviors with high resolution and repeatability, unveiling previously hidden nuances in animal interactions.

High-Throughput Behavioral Quantification

- **Pose Estimation and Tracking:** Sophisticated Deep Learning frameworks (like DeepLabCut or SLEAP) are trained to identify and track the precise location of an animal's body parts (e.g., nose, joints, limbs) across video frames. This allows researchers to analyze subtle movements, such as social cues, complex foraging techniques, or minute physiological responses to stim-

Isbn: 979-8-9950196-2-6

uli, providing a detailed, quantitative ethogram that was previously unattainable.

- **Automated Social Network Analysis:** By tracking the proximity and duration of interactions between tagged or identifiable individuals, ML algorithms can automatically map complex social networks within a population (e.g., primate troops or fish schools). This analysis is crucial for understanding disease transmission, cooperation, and the evolution of sociality.

The application of these tools effectively transforms behavioral data from qualitative observation into large-scale, quantifiable time-series data, suitable for rigorous statistical analysis.

3. Decoding Life: AI in Evolutionary and Molecular Zoology

At the molecular level, zoology is grappling with immense volumes of genetic data. AI and ML are essential for extracting biological meaning from these complex datasets.

Genomics and Taxonomy

- **Rapid Genomic Annotation:** ML algorithms accelerate the annotation of animal genomes by predicting the function of genes and non-coding regions, which is vital for understanding species adaptation and evolutionary divergence.
- **Integrative Taxonomy:** Species identification, particularly for cryptic species (those morphologically indis-

Isbn: 979-8-9950196-2-6

tinguishable), is being revolutionized. ML integrates traditional morphological measurements with modern molecular data (DNA barcoding, whole genomes) to provide highly accurate and objective taxonomic classifications, helping to map the tree of life more accurately.

Environmental DNA (eDNA)

AI/ML is critical for interpreting data from **eDNA**, the genetic material animals shed into the environment. ML models are used to filter out environmental "noise" and accurately identify the mixture of species present in a water or soil sample, offering a non-invasive snapshot of ecosystem biodiversity.

4. The Path Forward: Challenges and the Algorithmic Future

Despite its transformative power, the full potential of AI in zoology is constrained by several practical and ethical challenges.

- **Data Quality and Bias:** AI models are heavily reliant on the quality and representativeness of their training data. For rare species or underexplored behaviors, data scarcity is a major hurdle. Furthermore, biased training sets can lead to models that underperform for specific populations or environmental conditions.
- **Model Interpretability (Explainable AI):** For conservation decisions with high stakes, zoologists need to understand *why* an AI made a particular prediction. The "black box" nature of some deep learning models

Isbn: 979-8-9950196-2-6

presents a barrier to trust and deployment in critical management scenarios.

- **Accessibility and Collaboration:** The computational power and specialized programming expertise required to develop and deploy AI/ML models are not uniformly distributed. Future efforts must focus on building user-friendly, open-source tools and fostering collaborative platforms to ensure these technologies benefit researchers worldwide.

In conclusion, AI and ML are no longer futuristic concepts in zoology; they are the present-day engines of discovery. By moving beyond human-scale limitations in data processing, they are equipping zoologists to face the daunting challenges of the Anthropocene. The "Algorithmic Zoologist" promises a future where sophisticated computation is seamlessly integrated into every facet of animal study, leading to smarter conservation, deeper behavioral understanding, and a more complete map of life on Earth.

www.ingramcontent.com/pod-product-compliance
Lightning Source LLC
LaVergne TN
LVHW050917210926
848482LV00012B/469